Transition Tasks
for Mathematics
Grade 6

WALCH EDUCATION

1 2 3 4 5 6 7 8 9 10
ISBN 978-0-8251-6999-1

J. Weston Walch, Publisher
Portland, ME 04103
www.walch.com

Printed in the United States of America

Table of Contents

Introduction

Use these engaging problem-solving tasks to help transition your mathematics program to the knowledge and skills required by the Common Core State Standards for Mathematics.

This collection of tasks addresses some of the new, rigorous content found in the Common Core State Standards (CCSS) for sixth grade. The tasks support students in developing and using the Mathematical Practices that are a fundamental part of the CCSS. You can implement these tasks periodically throughout the school year to infuse any math program with the content and skills of the CCSS.

These tasks generally take 30 to 45 minutes and can be used to replace class work or guided practice during selected class periods. Depending on the background knowledge and structure of your class, however, the tasks could take less or more time. To aid with your planning, tasks are divided into two parts. This flexible structure allows you to differentiate according to your students' needs—some classes or advanced students may need only one class period for both parts, while others may need to defer Part 2 for another day or altogether. Use your own judgment regarding the amount of time your students will need to complete Parts 1 and 2. Another strategy for compressing the time necessary to complete a task is to divide the problems or calculation associated with a task among students or small groups of students. Then students can "pool" their information and proceed with solving the task.

Each Transition Task is set in a meaningful real-world context to engage student interest and reinforce the relevance of mathematics. Each is tightly aligned to a specific standard from the Grade 6 CCSS. The tasks provide Teacher Notes with Implementation Suggestions that include ideas for Introducing, Monitoring/Facilitating, and Debriefing the tasks in order to engage students in meaningful discourse. Debriefing the tasks helps students develop and enhance their understanding of important mathematics, as well as their reasoning and communication skills. The Teacher Notes also offer specific strategies for Differentiation, Technology Connections, and Recommended Resources to access online.

Student pages present the problem-solving tasks in familiar and intriguing contexts, and require collaboration, problem solving, reasoning, and communication. You may choose to assign the tasks with little scaffolding (by removing the sequence of steps/questions after the task), or with the series of "coaching" questions that currently follow each task to lead students through the important steps of the problem.

We developed these Transition Tasks at the request of math educators and with advice and feedback from math supervisors and middle-school math teachers. Please let us know how they work in your classroom. We'd love suggestions for improving the tasks, or topics and contexts for creating additional tasks. Visit us at www.walch.com, follow us on Twitter (@WalchEd), or e-mail suggestions to customerservice@walch.com.

Building Projects with Fractions

Common Core State Standard

Apply and extend previous understandings of multiplication and division to divide fractions by fractions.

6.NS.1. Interpret and compute quotients of fractions, and solve word problems involving division of fractions by fractions. ...

Task Overview

Background

This task uses division of fractions in the real-world application of building a deck railing and bookcase. It is designed to help students visualize the process of division by fractions while also providing skill practice.

Prior to this task, students should have learned and practiced addition, subtraction, multiplication, and division of fractions. They should have familiarity with reading a word problem and translating information into a sketch.

Students may have difficulty working with fractions. Division of fractions can be particularly challenging as it is often taught as a rote skill. This activity extends the division of fractions into the real world by asking, "How many _______ are in _______?"

The task also provides practice with:

- reading and interpreting word problems
- drawing diagrams to represent a problem
- performing arithmetic operations

Implementation Suggestions

- Students should work in small groups to encourage discourse.
- Each student should complete his or her own task.
- Calculator use is discouraged for this task.
- Students may benefit from the use of fraction tiles during this task.

Introduction

Ask students if they have ever built or helped build anything. Ask them to describe how they knew what to build. Ask if they used measurements, and if so, what kind. Did they use a model or sketch to represent the project? Was their model to scale? Ask students to explain why it is useful to have a sketch available when building an object.

Ask students to think about how much of a pizza they would have if they had to share half a pizza with three friends. Have students think silently and write/draw responses, then share out loud. Listen to the language used by the students. Encourage them to notice the connection between the words that describe sharing a pizza and the words used to express the operation of division.

Monitoring/Facilitating the Task

Ask questions and prompt student thinking so that they:

- Calculate the perimeter of the deck. They should label their sketch appropriately and review it as they calculate.
- Make a thoughtful choice about working with mixed numbers or improper fractions.
- Remember the procedure for dividing fractions.
- Think of the whole number or fraction being *divided*. Ask, "How many _______ are in _______?"
- Go deeper in their exploration when they calculate how many books are in a bookcase.
- Think about how the books would be placed on each shelf in order to encourage use of the proper dimensions.
- Describe how they came up with their answer and how they know they are correct.
- State and justify their solutions to one another.

Debriefing the Task

- Encourage students to present and describe the sketches they produced to represent the house and bookcase. Ask students to describe how they knew which labels to place where. Ask students to describe how the sketches were useful in their calculations.
- In Part 1, some students may have calculated the entire perimeter of the porch and then divided. Other students may have calculated the pickets on each side or section of the porch and found the sum for the house. Ask students to explore why both methods gave the same result.

- In Part 2, some students may use the mixed number as given while some students may have converted to improper fractions. Ask students which form helps them visualize the problem and which form made the division a simpler task.
- Encourage discussion about construction. Have students suggest additional items that need to be designed, remaining mindful of the thought, "How many _______ are in _______?"

Answer Key

Answers may vary depending on the batch combinations decided on by each team. Likely solutions are presented below.

1. Answers may vary but must include the house, deck, railing, and railing length.
2. 504 pickets. Students may suggest 505 pickets so that the railing is symmetrical on both sides. This answer is acceptable if the student provides an explanation.
3. Answers may vary but must include the top, sides, and correct number of shelves of the bookcase.
4. 7 shelves (*Note*: Students may calculate 8 shelves, but this would not account for the depth of the wood/shelves themselves.)
5. 4 bookcases—each bookcase holds 210 books, and $700 \div 210 \approx 3.33$. Round up to 4 to accommodate all the books.

Differentiation

Asking students to consider additional or different constraints on the problem will provide a variety of related tasks. Changing the dimensions of the house, bookcase, or books will result in new problems related to the initial assignment.

Advanced students can be asked to consider the dimensions of the wood they are using. Ask students to consider the number of pickets necessary if each corner contains a 4×4 post or if a supporting post must be placed at an interval around the railing. Ask students to consider the number of shelves or books if the wood being used is $\frac{3}{4}$-inch thick.

Technology Connection

Graphic design software could be used to "build" the railing and bookcase during the task.

Choices for Students

Students can determine the necessary pickets for a railing for their own porch, or a fence to surround the school or neighborhood park. Students may also design a bookcase for their own book collection or the library.

Meaningful Context

Fractions are everywhere in our world, yet many students have difficulty connecting classroom applications with real-world purposes. This task illustrates how fractions are commonly used in the construction trades.

Recommended Resources

- Bridge Builders
  www.walch.com/rr/CCTTG6FractionBridge
  In this online fraction activity, students must select the correct fraction to build a bridge with a predetermined number of sections.
- Dividing Fractions
  www.walch.com/rr/CCTTG6DivideFractions
  Review information on dividing fractions.
- How to Build a Bookcase
  www.walch.com/rr/CCTTG6BuildBookcase
  This site provides background information for the teacher on building a bookcase.
- How to Build a Porch Railing
  www.walch.com/rr/CCTTG6PorchRailing
  This site provides background information on building a porch railing.

NAME:

6.NS.1 Task • The Number System
Building Projects with Fractions

Part 1

Your client wants your construction company to build a railing along her deck. She wants a basic picket railing like the one in the following diagram. How many pickets will you need?

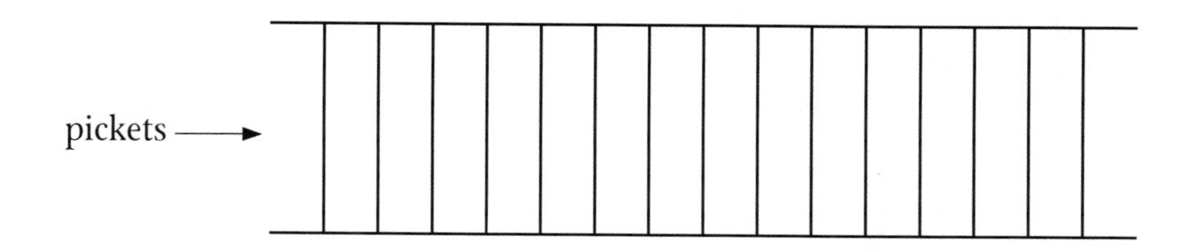

You have collected the following information:

- The client's house is a square.
- Each side of the client's house is 10 feet long.
- The deck surrounds 3 sides of the client's house.
- The deck extends 9 feet from the house.

1. Sketch the aerial view of the house. Include the house, deck, and railing. Make sure to label all information and known measurements.

2. Calculate how many pickets are needed for the railing. Each picket must be spaced $\frac{1}{6}$ foot apart.

continued

6.NS.1 Task • The Number System
Building Projects with Fractions

Part 2

Your company did such excellent work on the railing that your client wants you to build a bookcase. Your client has a list of requirements. How many bookcases will you need to build to hold 700 books?

- The bookcase must be 6 feet tall.
- The bookcase must be 4 feet wide, which includes the two wooden sides of the bookcase. The wood on each side is $1\frac{1}{2}$ inches thick, leaving $3\frac{3}{4}$ feet of space on the shelf for books.
- The bookcase needs to hold books that are $\frac{3}{4}$ foot tall and $\frac{1}{8}$ foot wide.
- Shelves must be evenly spaced.
- Shelves must be 1 inch thick. The top and bottom of the bookcase are also 1 inch thick.

3. Sketch the bookcase. Include the top, sides, and shelves. Make sure to label all information and known measurements.

4. Calculate how many shelves the bookcase will hold. Correct your sketch if necessary.

5. Your client owns 700 books. How many bookcases do you need to build? Justify your response.

Port Royal: A Sunken City

Common Core State Standard

Apply and extend previous understandings of numbers to the system of rational numbers.

6.NS.6b. Understand signs of numbers in ordered pairs as indicating locations in quadrants of the coordinate plane; recognize that when two ordered pairs differ only by signs, the locations of the points are related by reflections across one or both axes.

Task Overview

Background

Students sometimes struggle to develop the understanding that signed numbers in ordered pairs indicate specific locations in the quadrants of a coordinate plane. This is often revealed when they are asked to identify locations within specific quadrants or when asked to plot ordered pairs that differ only by their signs. It is important for students to recognize that when ordered pairs differ only by their signs, they are a reflection across one or both of the axes. Students need multiple opportunities to practice plotting ordered pairs and identifying the quadrants. This task helps students use a real-world situation to enhance their understanding of the location of ordered pairs as well as reflections of ordered pairs across axes.

The task also provides practice with:

- identifying x- and y-coordinates
- explaining reasoning

Implementation Suggestions

Students should work individually to complete the activity and should then meet in pairs or in small groups to compare and discuss their work before a class-wide discussion. Students may want to use a colored pencil or marker when plotting on the coordinate plane.

Introduction

Introduce the task by asking students about their experience with ordered pairs. Ask them for examples of ordered pairs used in real-world situations. Ask students if they are familiar with underwater archaeology and how the use of coordinate planes and ordered pairs may be helpful in this context.

Monitoring/Facilitating the Task

Ask questions and prompt student thinking so that they:

- Correctly plot each of the given ordered pairs.
- Correctly identify each of the quadrants.
- Recognize the signs of each value of the coordinate pairs as it relates to the quadrants.
- Understand reflection transformations.
- Distinguish the x-axis from the y-axis.

Debriefing the Task

- Circulate to ensure that students have correctly plotted the ordered pairs before students discuss.
- Make sure students discuss their reasoning behind the number of rooms the building contained. Students should mention the location and number of doorways as well as staircases in their reasoning.
- Facilitate a brief class discussion about how students determined the number of rooms within the building. Ask students what role the coordinate plane had in making their determination. Students may mention the reflection of the staircases over the y-axis as well as the translation or shift from room to room.
- Students should understand that the coordinate plane is a map of the underwater city and a way to record the location of the findings.

Answer Key

Students' completed graphs should contain the following points:

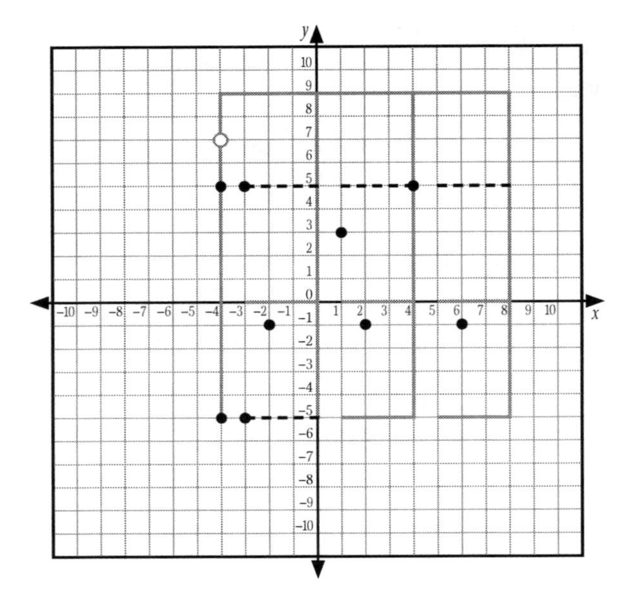

1. (2, –1)
2. (–3, –5) and (–4, –5)
3. (4, 5)
4. (6, –1); Explanations may vary, but should include a reference to the third set of stairs being located 4 units to the right of the stairway found in problem 1. All three sets of stairs should have the same y-value, since the stairs are straight.
5. Quadrant I. Both the x- and y-values are positive. All ordered pairs in Quadrant I have positive x- and y-values.
6. Answers may vary, but should include a negative x-value and a negative y-value. Ordered pairs in Quadrant III are always of the form (negative, negative).
7. Quadrant II. The ordered pair (–4, 7) has a negative x-value and a positive y-value, like all coordinates of Quadrant II.
8. Answers may vary, but responses should include references to the signs of the coordinates. Artifacts with coordinates in the general form of (x, y) are located in Quadrant I; artifacts with coordinates of $(-x, y)$ are located in Quadrant II; artifacts with coordinates of $(-x, -y)$ are located in Quadrant III; and artifacts with coordinates of $(x, -y)$ are located in Quadrant IV.

Differentiation

Some students may benefit from labeling each of the quadrants prior to beginning the task. Students who finish the task early can create a map of a fictitious underwater city and write questions similar to those found in the task.

Technology Connection

Students can use graphing software such as GeoGebra to plot the points used in the task. Students could also use computer-assisted drawing software to create a final blueprint of the building.

Choices for Students

After reading through the task, students could create a scavenger hunt based upon using a coordinate plane as a map. Questions/clues should be based on the reflection of ordered pairs over one (or both) axes. References to quadrants should also be included.

Meaningful Context

Underwater archaeologists continue to investigate the ocean, not just to find treasures of the past, but also to learn from and about history. In order to relocate findings, it is important to create a detailed drawing that includes precise locations. One tool used to record the location of objects is a coordinate plane. Having an understanding of ordered pairs and their relation to location is essential in this field of study.

Recommended Resources

- Graphing Equations and Inequalities: The Coordinate Plane
  www.walch.com/rr/CCTTG6CoordinatePlane
  This site provides a brief overview of graphing ordered pairs, complete with labeled drawings.
- Interactivate: Maze Game
  www.walch.com/rr/CCTTG6MazeGame
  This applet allows students to explore ordered pairs while trying to avoid mines within a maze.
- The Port Royal Project
  www.walch.com/rr/CCTTG6PortRoyal
  Review the history of Jamaica's Port Royal along with details of the archaeological excavations that have taken place there.

NAME:

6.NS.6(b) Task • The Number System
Port Royal: A Sunken City

Introduction

The population of Port Royal, a city in southeastern Jamaica, grew so quickly that by the mid-seventeenth century, people were filling in areas of water in order to create more land for building new homes and stores. On June 7, 1692, an earthquake caused much of the sandy city to sink straight down into the ocean. Since then, underwater archeologists have explored the sunken ruins. They have recorded their findings of one particular building on a coordinate plane. How could the use of coordinates help archaeologists determine which quadrant an artifact is located in? Use the drawing on the coordinate plane below to record your answers to the following questions.

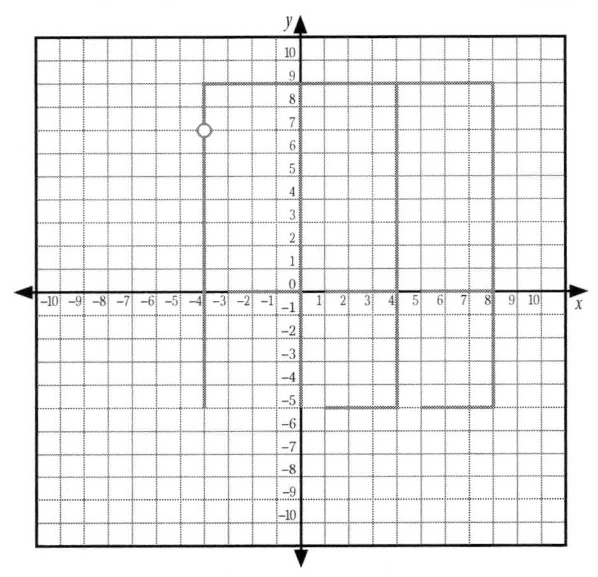

Task Questions

1. Archaeologists found evidence of a stairway located at the ordered pair (–2, –1). Another stairway was found at the ordered pair represented by the reflection of (–2, –1) over the y-axis. What ordered pair represents the second stairway? Plot both ordered pairs on the coordinate plane.

2. Evidence of a doorway was found at (–3, 5) and (–4, 5). A similar doorway was found at the ordered pairs that are a reflection of (–3, 5) and (–4, 5) over the x-axis. What ordered pair represents this doorway? Plot the ordered pairs that represent both doorways on the coordinate plane.

3. An entry was found at the location of the reflection of (–4, –5) over both the x- and y-axes. What is the location of this entry? Plot the ordered pair on the coordinate plane.

4. It is believed a third set of stairs was located in the fourth quadrant. Based on information that you've already been given about the two other stairways, where do you think the third set of stairs was located? Plot the ordered pair. Explain your answer.

5. Scraps of leather were found at (1, 3). In which quadrant was the leather found? Plot the location of the leather. Explain why you plotted the point in this quadrant.

6. Soles of shoes were found inside the building in the third quadrant. Plot a possible location of a shoe sole. Explain your answer.

continued

6.NS.6(b) Task • The Number System
Port Royal: A Sunken City

7. A water storage tank was found located on the exterior wall at the point (–4, 7). In which quadrant is the tank located? Explain your answer.

8. Based on what you have found, how would you determine in which quadrant an artifact lies given its coordinates?

Golfing with Number Lines and Coordinate Planes

Common Core State Standard

Apply and extend previous understandings of numbers to the system of rational numbers.

6.NS.6c. Find and position integers and other rational numbers on a horizontal or vertical number line diagram; find and position pairs of integers and other rational numbers on a coordinate plane.

Task Overview

Background

Students may have difficulty understanding negative integers and how they compare to positive integers. This task reinforces the skill of plotting positive and negative integers and decimals on a number line and extends it to comparing positive and negative integers on the number line. Students then enhance their understanding of integers on a number line by identifying and plotting ordered pairs on a coordinate plane.

The task also provides practice with:

- comparing positive and negative numbers
- translating points on a coordinate plane

Implementation Suggestions

Students may work individually or in pairs to complete one or both parts of the task. Alternatively, students may complete each part of the task individually and then meet in groups to share their results before a class-wide discussion.

The task should take 20 minutes for Part 1 and 10 minutes for Part 2. The debrief will take an additional 20 minutes. Display a number line spanning from –10 to 10 in increments of 1 and a Cartesian coordinate plane with axes labeled from –10 to 10 during the debrief. (These are provided for you following the Recommended Resources if you wish to use an opaque projector or copy them onto a transparency.)

Introduction

Introduce the task by asking students if they have ever played golf. Have any of them heard the term *par* used on the golf course? Explain that, in golf, each hole has a number of strokes (or swings with the golf club) that should be required to complete the hole. This number is called par. If par is 5 and it takes you 6 strokes to complete the hole, your score is +1, or 1 over par. If par is 5 and it takes you 4 strokes to complete the hole, your score is –1, or 1 under par. The scores for each hole are added up to find your score for a round (18 holes). Contrary to scoring rules for other sports, the best golfers have the lowest scores. Ask students if they have ever kept track of scores for a game or a sport similar to golf.

Monitoring/Facilitating the Task

Ask questions and prompt student thinking so that they:

- Represent and compare numbers on a number line. Make sure that students articulate how to compare positive and negative numbers on the number line.
- Place decimals correctly on a number line. Ensure students articulate how to plot a number that is between two whole numbers.
- Understand how points can be plotted on a coordinate plane. If students are having difficulty explaining their responses, ask them to explain verbally how they determined their answers. Be sure students understand how to properly write coordinates of a point.
- Prompt for and encourage the use of proper mathematic terms as well as justification of strategies and solutions.

Debriefing the Task

- Ensure students recognized how to plot and compare numbers and how to identify and plot coordinates on a coordinate plane.
- Prompt students to explain their thinking about plotting positive and negative integers on the number line.
- Ask students to describe how they can determine which number on a number line is highest or lowest. State a number between –10 and 10 and ask students to compare the number with those plotted on the number line.
- Encourage students to explain how they plotted the decimal numbers. State positive and negative decimal numbers and have students state which two whole numbers the decimal would be between on the number line.

- Ask students to share their coordinate planes with all the points plotted. Discuss with students the importance of the order of the coordinates in an ordered pair.
- Point to each quadrant and ask students whether the first coordinate of the ordered pair will be positive or negative, and whether the second coordinate of the ordered pair will be positive or negative. Guide students to state a rule for the signs of the coordinates for each quadrant.
- Prompt students to discuss how points on a coordinate plane can be translated. Encourage students to share the steps they took to translate a point.
- Assess understanding by expanding the parameters used in the task. Have students compare the locations of all the points plotted on the coordinate plane. Ask which chip was closest to the hole and which chip was farthest from the hole. Discuss alternate ways to find the answer, such as subtracting from or adding to the *x*- and *y*-coordinates of the original point.
- Discuss student experiences with keeping track of sports results or game points. Ask students to explore how they could use a number line or a coordinate plane to record their own results.

Answer Key

1. It would be to the right of Tyra's score because it is a positive number and Tyra's score is a negative number.
2. Check student number lines for accuracy—the points 6, 7, 8, and 9 should be plotted. The score 9 is the highest. You can tell this on the number line because it is farthest to the right.
3. Check student number lines for accuracy—the points –2, –1, 0, and 2 should be plotted. The score –2 is the lowest. You can tell this on the number line because it is farthest to the left.
4. Check student number lines for accuracy—the points 7.5 and –0.25 should be plotted. The point 7.5 is halfway between 7 and 8. The point –0.25 is to the left of 0, one-fourth of the way between 0 and –1.
5. Completed number line:

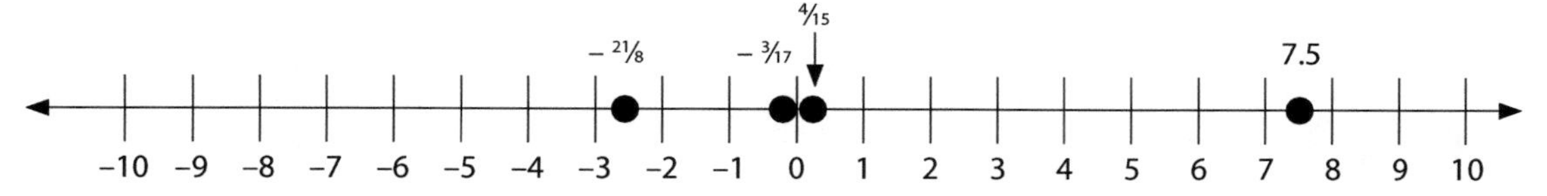

   Each of my friends' average scores is lower than my average score; they are all to the left of my score on the number line.

6. (–4, 6)
7. Check student graphs. The point (2, 3) should be plotted on the coordinate plane.

8. Check student graphs. The point (3.5, 5.5) should be plotted on the coordinate plane.
9. Check student graphs. The point (6.4, –7.25) should be plotted on the coordinate plane.
10. The coordinates are (–6 1/3, 9). I moved 2 1/3 units to the left of the first point, and then I moved 3 units up.

Differentiation

Some students may benefit from working with a partner to complete the tasks. Encourage students who complete Part 1 early to find the difference between their average score and Tyra's average score. Have students explain how they can use the number line to solve this problem. Students who complete Part 2 early can answer the following question: In which quadrant of the coordinate plane would the fourth ball be chipped in if it landed to the right of the hole and below the hole on the coordinate plane? Have students explain their answer. Students who complete the task early may also work with a blank coordinate plane to create another scenario (e.g., pinpointing animal burrows in a field) in which locations can be indicated through the use of coordinate pairs.

Technology Connection

Students could use graphing software to plot the points. If appropriate equipment is available, students could use the Internet to look up information on specific sports results. For example, students could look up golf scores for the latest U.S. Open or Masters Tournament, the number of points scored by each team in the Super Bowl, or world records for athletic events. Students could use software to create a report or poster to summarize the information.

Choices for Students

Following the introduction, offer students the opportunity to choose a sport or game that they enjoy. Have them develop a way to use number lines or coordinate planes to record a result of some type from the game or sport.

Meaningful Context

This task may be expanded by having students record and plot data for a sport or game that they play. Have students select data to collect. Students should record the data and then plot it on either a number line or a coordinate plane. Ask students to explain how recording the data can provide them with information to help them improve in their sport or game.

Recommended Resources

- Coordinate Geometry Activities
  www.walch.com/rr/CCTTG6GeometryGames
  This site features a number of games that introduce and use coordinate geometry. Paired or in small groups, students can learn how to identify coordinates in the first quadrant. Download directions and game mats for each game directly from the site. Some extra game materials (dice, small cubes, etc.) are required.
- Line Jumper
  www.walch.com/rr/CCTTG6LineJumper
  This number line game gives students practice in locating numbers on a number line and using number lines to complete addition and subtraction problems. The game has a range of difficulties to choose from, and is suitable as a revision or an extension task.
- Locate the Aliens—Name the Coordinates of a Point on a Graph
  www.walch.com/rr/CCTTG6LocateAliens
  This game allows students to practice finding coordinates by locating aliens on a coordinate plane. The game helps students solidify their ability to identify points with positive and negative number coordinates.

Cartesian Plane and Number Line

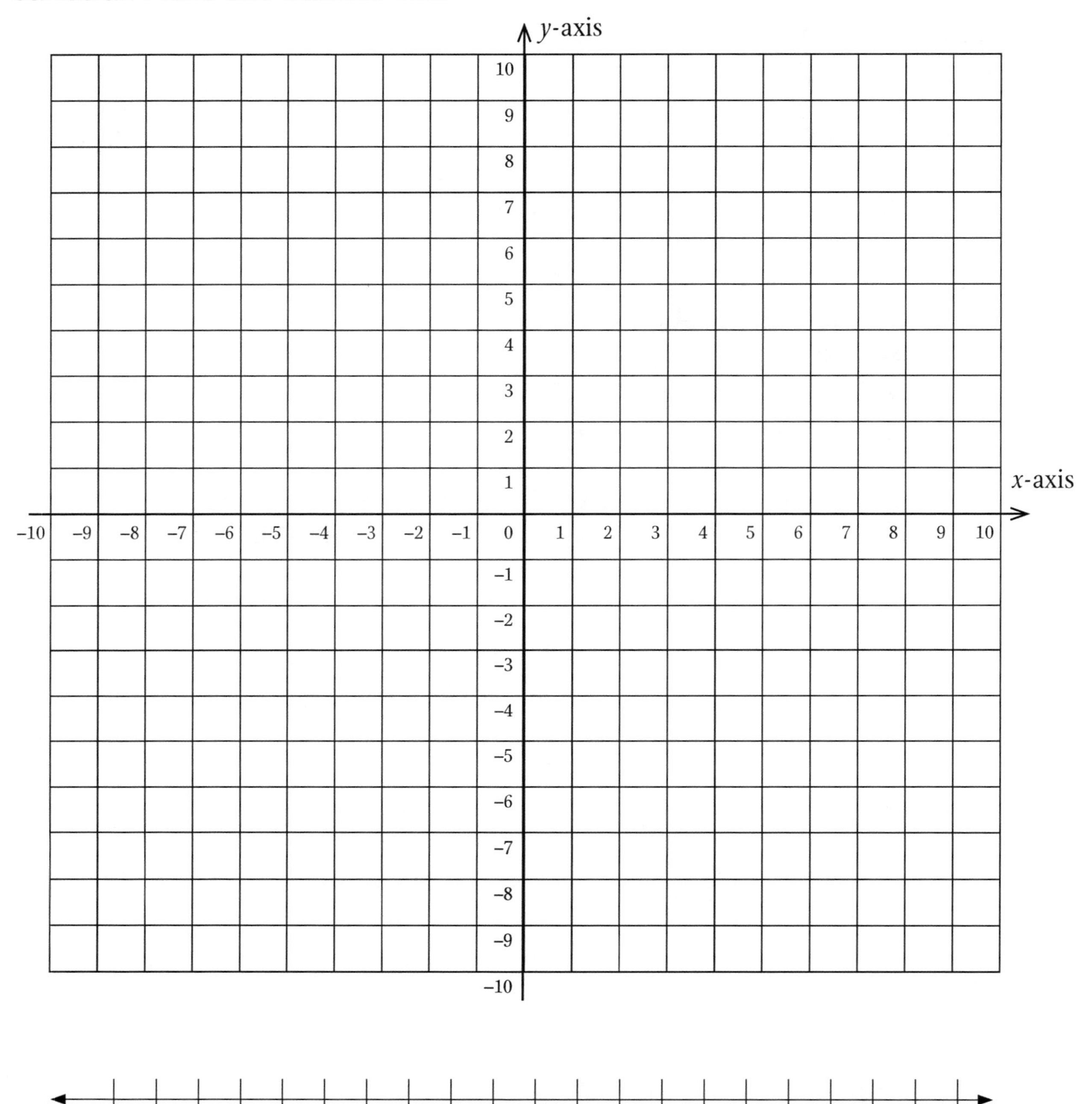

–10 –9 –8 –7 –6 –5 –4 –3 –2 –1 0 1 2 3 4 5 6 7 8 9 10

NAME:

6.NS.6(c) Task • The Number System
Golfing with Number Lines and Coordinate Planes

Part 1

You are practicing for a golf competition. Over the weekend, you play four rounds of golf with your coach, Tyra. You record the scores after each round, and then find the average of the four scores. How can you use a number line to compare scores? Show some examples and explain your procedure.

	Your score	Tyra's score
Round 1	+8	–2
Round 2	+7	–1
Round 3	+9	+2
Round 4	+6	+0
Average score	+7.5	–0.25

1. If the scores for Round 1 were plotted on a number line, would your average score be to the right or to the left of Tyra's average score? Explain your answer.

2. Plot your scores for the four rounds on the number line. Which score is the highest? Explain your answer.

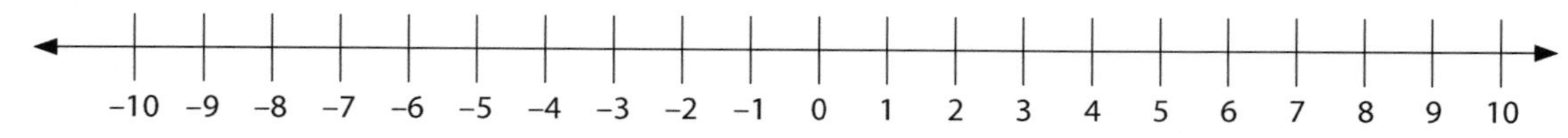

3. Plot Tyra's scores for the four rounds on the number line. Which score is the lowest? Explain your answer.

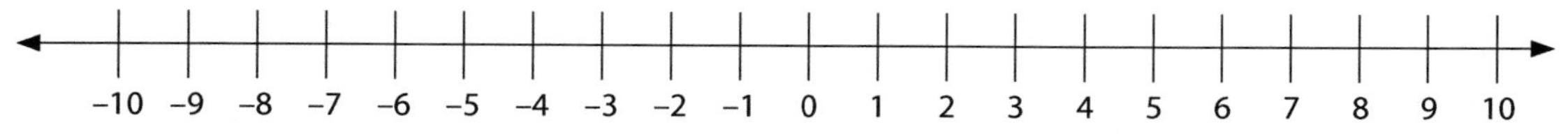

continued

6.NS.6(c) Task • The Number System
Golfing with Number Lines and Coordinate Planes

4. Plot your average score and Tyra's average score on the number line. Explain how you determined where to plot each point.

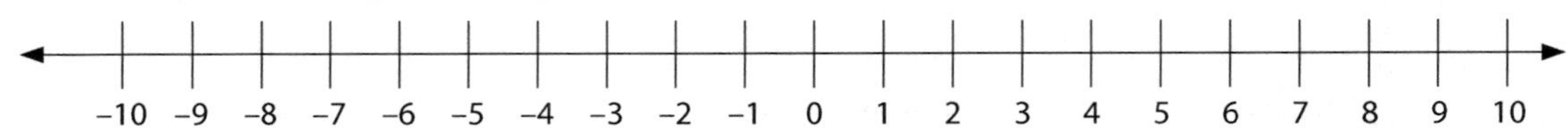

5. You want to know how your average compares with everyone else's, so you ask three of your friends for their average scores so far this season. Jean tells you his average score is $\frac{4}{15}$; Reggie tells you his average is $-\frac{3}{17}$; and Denise tells you her average is $-\frac{21}{8}$. Plot your average score along with the average score for each of your friends on the number line below. How does your score compare to your friends' scores?

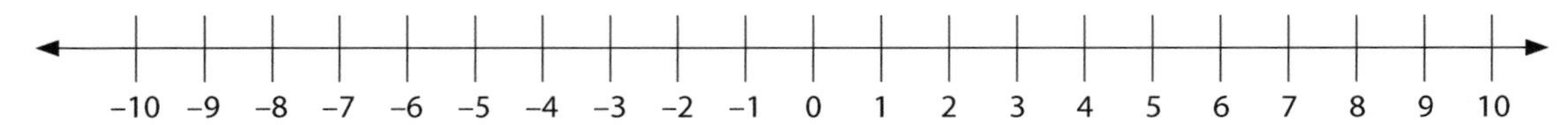

6.NS.6(c) Task • The Number System
Golfing with Number Lines and Coordinate Planes

Part 2

You are practicing chipping the golf ball onto the green (the area of grass around the hole). Chipping means using the golf club to pop the ball into the air a short distance so that it lands on the green. You want to record where the ball lands. You draw a coordinate plane to represent the green. The point (0, 0) represents the hole, and each unit on the plane represents 1 foot. You measure how far each ball is from the hole. The location of the first ball you chip is plotted below.

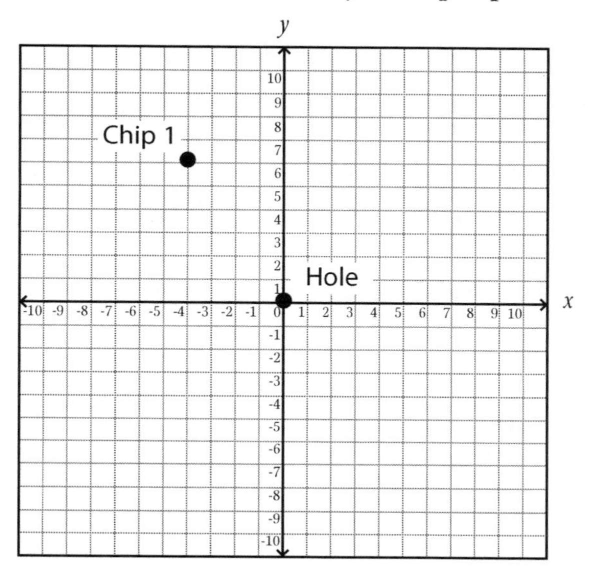

6. What are the coordinates of the first ball you chip?

7. The second ball you chip lands at (2, 3). Plot this point on the coordinate plane.

8. The third ball you chip lands at (3.5, 5.5). Plot this point on the coordinate plane.

continued

9. The fourth ball you chip lands $\frac{32}{5}$ feet to the right of the hole and $-\frac{29}{4}$ feet below the hole on the coordinate plane. Plot this point on the coordinate plane.

10. The last ball you chip lands $\frac{7}{3}$ feet to the left of the first ball and 3 feet above it on the coordinate plane. What are the coordinates of the point where the last ball lands? Explain how you found the coordinates.

Who Ordered This Weather?

Common Core State Standard

Apply and extend previous understandings of numbers to the system of rational numbers.

6.NS.7b. Write, interpret, and explain statements of order for rational numbers in real-world contexts. For example, write –3°C > –7°C to express the fact that –3°C is warmer than –7°C.

Task Overview

Background

Students often struggle when ordering rational numbers, especially when fractions and negative values are involved. This task helps students internalize rational numbers and ordering skills, while focusing on temperatures and precipitation amounts.

The task also provides practice with:

- creating equivalent fractions
- converting fractions to decimals
- reading data tables
- writing algebraic expressions as verbal representations
- writing inequalities from verbal phrases

Implementation Suggestions

Students may work individually or in pairs to complete one or both parts of the task. Alternatively, students may meet in groups to share their results and reflect after individually completing the task, before a class-wide discussion. Have number lines available for use, if necessary. Review Celsius vs. Fahrenheit if necessary.

Introduction

Introduce the task by asking students if they have ever researched monthly temperatures or precipitation amounts for a specific area. (Students are likely to have some experience with this, as researching temperature and precipitation is a fairly common activity in science, social studies, and/or math classes in the elementary grades.) Ask students how these values are determined, as well as who might want to know the average temperature and precipitation amounts and why. Together, create a list of occupations and interests that are weather dependent (e.g., farmers, oceanographers, bird watchers, beekeepers). Continue by asking students if they are aware of trends in the weather or can recall periods of abnormal weather.

Monitoring/Facilitating the Task

Ask questions and prompt student thinking so that they:

- Recognize the mathematical procedures they are following.
- Articulate how they are ordering and comparing rational numbers. *Note:* If a student is struggling, redirect them to a number line. Allow students to place each number on the number line and then answer the questions.
- Clearly articulate how they used their calculations to answer each question.
- Avoid relying on ideas such as "the alligator eats the bigger number," but can read each sentence using the correct mathematical terms.
- Recognize and acknowledge that there are several ways to compare numbers; students may choose to create equivalent fractions or convert fractions to decimals.

Debriefing the Task

- Students will be comparing rational values and must recognize how to appropriately order rational numbers.
- Problem 1 requires that students write an algebraic statement that compares two values. Choose two students with different correct answers to share their responses to ensure students see both approaches. Discuss with students why both answers are correct. If students have trouble with this concept, you may want to follow up with a similar example.
- For question 2, verify that students understand the statement by having them explain their thought processes.
- In order to compare fractions, students could choose to convert the fractions to decimals or to create equivalent fractions. Select students who used different processes to explain their answers. Students should recognize that the processes result in the same answer.
- Discuss the phrase, "it's too cold to snow," and why this expression is used. Ask students if they have ever heard this expression. Have students defend their answers to question 6 based on the data provided in the table.
- Students may struggle with writing a statement for problem 7 due to the use of the variable R for the concrete temperature, and may try to incorporate T and C into their statements. Encourage students to use only the specific temperature values and the variable R.
- Give students a value not listed in the table. Ask them to determine whether the value fits their statement in problem 8.
- Encourage students to share their arithmetic-based or logical reasoning methods for their responses to question 10.

Who Ordered This Weather?

Answer Key

1. $-4.17°C < -2.61°C$ or $-2.61°C > -4.17°C$
2. December is warmer than January. I know this because –1.39°C is greater than –4.17°C.
3. February has the greater average precipitation amount because $3\,^{57}/_{100}$ is greater than $3\,^{1}/_{10}$.
4. $3\,^{4}/_{25} > 3\,^{57}/_{100}$ or $3\,^{57}/_{100} > 3\,^{4}/_{25}$
5. $3\,^{1}/_{10} < 3\,^{4}/_{25} < 3\,^{57}/_{100}$
6. Answers may vary, but the evidence from the table suggests the phrase is incorrect, since January (the coldest month) has the most precipitation.
7. $R \geq 18.33°C$
8. When the air temperature is greater than or equal to –17.78°C, but less than or equal to –1.11°C, then the concrete must be at least 15.56°C.
9. 12.78°C
10. Generally, no, since the average temperature never falls below –17.78°C. However, some students may argue that during a particularly cold winter, some daily temperatures may fall well below the average temperature, and possibly into the –17.78°C range. Since the minimum concrete temperature depends on the daily air temperature and not the average monthly temperature, accept this answer given appropriate student justification.

Differentiation

Some students may benefit from the use of number lines during this task.

Students who finish early could research the average temperature and precipitation amounts for their town or city or another location of their choice. They could also research occupations that are dependent on the weather, as well as the impact of weather on those professions.

Technology Connection

Students could use a graphing calculator to sort the values in order to check if their answers are correct.

Choices for Students

Following the introduction, offer students the opportunity to choose their own location. Encourage students to use locations that have a variety of temperatures in order to compare negative rational numbers. Have them create their own statements, both verbal and algebraic, related to the information.

Meaningful Context

This task may be expanded by researching how temperatures or precipitation amounts have an effect on issues other than pouring concrete for construction. Students could also research the climatology and the impact, if any, of unusual high/low temperatures or precipitation amounts on the average values reported.

Recommended Resources

- Compare and Order Rational Numbers
  www.walch.com/rr/CCTTG6CompareOrder
  This site provides a quick reference for comparing and ordering rational numbers. On the right side of the page under "Other Resources," you will find links to practice problems and a video.
- Comparing and Ordering Rational Numbers Game
  www.walch.com/rr/CCTTG6GameShow
  This multiple-choice, interactive game provides practice with comparing rational numbers in a game show format that accommodates single players or up to four teams.
- Number Balls Game
  www.walch.com/rr/CCTTG6NumbersGame
  Users try to beat their high score by arranging floating, spinning integers in ascending order. Difficulty increases with each round.

6.NS.7(b) Task • The Number System
Who Ordered This Weather?

Part 1

Building contractors often plan their projects around the weather forecast in order to avoid extremely cold temperatures or heavy precipitation. It's important for them to be familiar with the local weather data. The table below shows the average temperature and the average precipitation totals for each of the winter months (December, January, and February) in Frostburg, Maryland. Analyze the information in the table to help identify the best times, during the winter, to schedule construction in Frostburg.

Winter Weather Data for Frostburg, Maryland

	December	January	February
Average temperature (°Celsius)	–1.39	–4.17	–2.61
Average precipitation totals (inches)	$3\frac{4}{25}$	$3\frac{57}{100}$	$3\frac{1}{10}$

Source: NOAA: Climatography of the United States for the Years 1971–2000; www.walch.com/CCTTG6USClimateData

1. The average monthly temperature for January is typically colder than the average monthly temperature for February. Compare the average temperatures for January and February using < or >.

2. The relationship between the average monthly temperatures for December and January can be expressed as $-1.39°C > -4.17°C$. Is the month of December warmer or cooler than the month of January? Explain your reasoning.

3. The relationship between the average precipitation totals for January and February can be expressed as $3\frac{57}{100}$ inches $> 3\frac{1}{10}$ inches. Which of the two months has the greater average total precipitation? Explain your answer.

6.NS.7(b) Task • The Number System
Who Ordered This Weather?

4. Write a mathematical expression that uses the symbol < or > to compare the average total precipitation amounts for the months of December and January.

5. Arrange the average total precipitation amounts for the months of December, January, and February from smallest to largest. Use the format given below.

6. The expression, "it's too cold too snow" is often used during the winter months. Based on the data in the table, do you agree or disagree with this statement? Explain your answer.

6.NS.7(b) Task • The Number System
Who Ordered This Weather?

Part 2

Contractors must continue working and pouring concrete through the winter months, even when the temperatures are cold. However, they have to take precautions that are not taken during warmer months. The minimum temperature of the concrete when poured depends on the air temperature. If the concrete is not warm enough, the concrete will set too quickly and result in a weak surface. In order to keep their building projects safe, contractors must be aware of the air temperature as well as the forecast while planning construction. The table below shows the recommended air temperature ranges for various concrete temperatures.

Air temperature (°C)	Minimum concrete temperature (°C)
$T > -1.11$	12.78
$-17.78 \leq T \leq -1.11$	15.56
$T < -17.78$	18.33

7. Write an inequality to describe the phrase, "the concrete temperature is at least 18.33°C." Let R stand for "concrete temperature."

8. According to the table, the concrete must have a minimum temperature of 15.56°C when $-17.78 \leq T \leq -1.11$. Use words to describe this situation.

9. On average, if a contractor is pouring concrete in Frostburg in January, what should the minimum temperature of the concrete be? *Hint:* Refer to the table in Part 1 to determine Frostburg's average January temperature.

10. Using the data from the table in Part 1, would you expect the concrete to ever have a minimum temperature of 18.33°C when building in Frostburg? Explain your answer.

Water Crisis in Haiti

Common Core State Standard

Apply and extend previous understandings of numbers to the system of rational numbers.

6.NS.8. Solve real-world and mathematical problems by graphing points in all four quadrants of the coordinate plane. Include use of coordinates and absolute value to find distances between points with the same first coordinate or the same second coordinate.

Task Overview

Background

Students often struggle when plotting coordinates in all four quadrants even after successfully plotting points in the first quadrant. Students sometimes confuse the directions of the axes and the coordinates that make up the point. This task allows students to plot points in a meaningful context—by focusing on solving water contamination and cholera outbreak problems caused by the 7.0 magnitude earthquake that struck Haiti on January 12, 2010.

The task also provides practice with:

- finding the horizontal or vertical distance between two points
- determining the absolute value of a number

Implementation Suggestions

Students may work individually, in pairs, or in small groups to complete one or both parts of the task. Part 1 should take approximately 10 minutes, and Part 2 should take approximately 20 minutes followed by a discussion/debrief for 20 to 30 minutes. *Please note*: If time becomes a concern, problem 10 can be made optional.

Students should be provided with a red pencil, marker, or pen to complete the task.

Introduction

Introduce the task by asking students about their knowledge of recent natural disasters, such as earthquakes, hurricanes, tornados, tsunamis, etc. Question students about their understanding of the health risks that ensue after a natural disaster. Ask students if they have ever lost their access to water following a storm and what they did to obtain clean water. Ask students if they have ever heard of cholera.

Monitoring/Facilitating the Task

Ask questions and prompt student thinking so that they:

- Understand and internalize the procedure for plotting points in four quadrants.
- Defend their responses. Make sure that students articulate how they used their calculations to answer a question.
- Recognize the arithmetic they are doing. If students are having difficulty describing their arithmetic or explaining their responses in writing, ask them to explain verbally how they determined their answers. Prompt for and encourage the use of proper mathematic terms.
- Realize that there are at least three ways to calculate distance: by using an algorithm, by using absolute value, and by counting.

Debriefing the Task

- Students will be plotting points in all four quadrants. They must recognize which coordinate belongs to which axes in the ordered pair.
- Ask students to explain how they plotted their points. Some students will choose to work with the x-coordinate first, while others may choose to work with the y-coordinate first. Try to find students who used each method and ask the class which is correct. Students should realize that both ways are correct.
- Students will be asked to find the distance between affected areas. Each of these distances makes use of ordered pairs that have either the same x-coordinate or the same y-coordinate.
- In problem 5, students will have to use absolute value to determine the distance if they use an equation. Prompt students to explain how they used the absolute value.
- Prompt students to think about why distance cannot be reported using negative numbers and ask them how absolute value can help with this.
- Encourage students to think about multiple methods for determining the distance between two points. Students might choose to count, use an algorithm, or make use of absolute value. Ask students to report their methods and be sure to have each method presented by at least one student.
- Assess understanding by expanding the parameters used in the task. Encourage students to plot more affected areas on the map and calculate distances.

Answer Key

1.

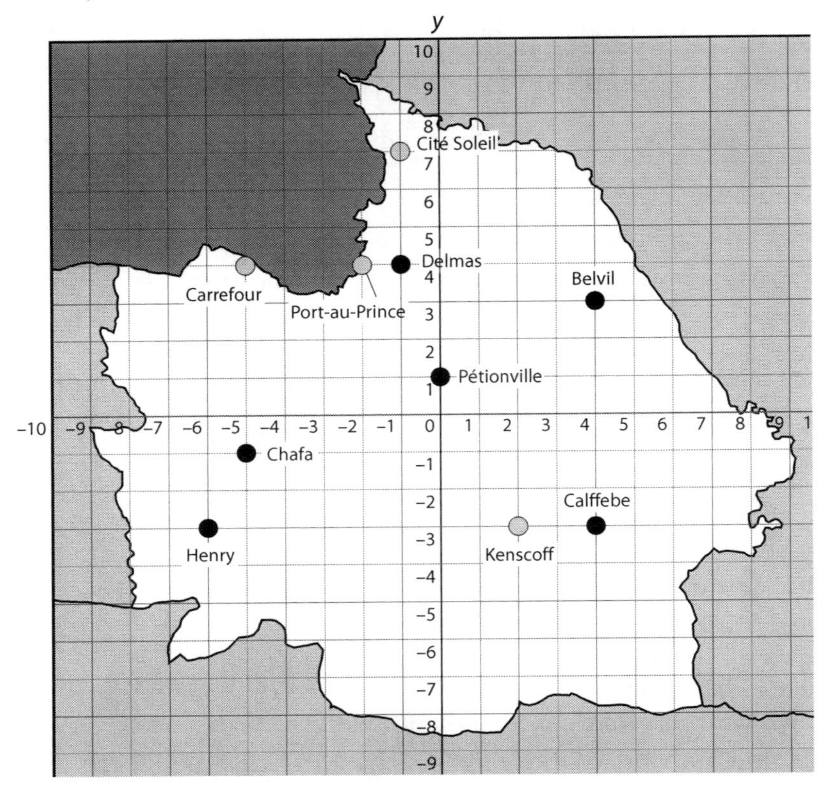

Affected area	Coordinates	Approximate reported cholera cases
Belvil	(4, 3)	4,500
Calffebe	(4, –3)	6,600
Carrefour	(–5, 4)	10,700
Chafa	(–5, –1)	4,000
Cité Soleil	(–1, 7)	11,100
Delmas	(–1, 4)	5,600
Henry	(–6, –3)	3,700
Kenscoff	(2, –3)	10,300
Pétionville	(0, 1)	3,000
Port-au-Prince	(–2, 4)	12,000

2. Check to see that Port-au-Prince (–2, 4), Kenscoff (2, –3), Cité Soleil (–1, 7), and Carrefour (–5, 4) are all marked in red.

3. Quadrant II, because this quadrant has the most affected areas with more than 10,000 cholera outbreaks.

4. 10 units; Henry is located at (–6, –3) and Calffebe is located at (4, –3). They have the same y-coordinate. Therefore, examine the x-coordinates and add their absolute values: $|-6| + |4| = 6 + 4 = 10$. Distance is always positive and the absolute value of a number is always positive. Absolute value is the distance from 0. This means that Henry is 6 units to the left of 0 and Calffebe is 4 units to the right of 0. Their sum is 10.

5. Check to see that Sarthe is plotted at (0, 7) and labeled.

6. Cité Soleil; (–1, 7); 1 unit; answers may vary. Sample answers: I counted the number of tick marks between Sarthe and Cité Soleil, or used the x-coordinates to create and solve the equation $0 - (-1) = 1$, or took the absolute value of $|-1|$.

7. 3; answers may vary. Sample answers: I counted the number of tick marks between Cité Soleil and Delmas, or used the y-coordinates to create and solve the equation $7 - 4 = 3$.

8. 3; answers may vary. Sample answers: I counted the number of tick marks between Port-au-Prince and Carrefour, or used the x-coordinates and absolute value to create and solve the equation $|-2 - (-5)| = 3$ or $-2 - (-5) = 3$ or $-5 - (-2) = |-3| = 3$.

9. Answers may vary. Be sure that students defend their responses. Sample answer: Chafa, because it is only 5 vertical map units and 0 horizontal map units away from the last affected area visited.

10. Answers may vary. Be sure that distances are correct. Sample answer:

Originating area	Horizontal distance (x-axis)	Vertical distance (y-axis)	Destination
Chafa	1	2	Henry
Henry	8	0	Kenscoff
Kenscoff	2	0	Calffebe
Calffebe	4	4	Pétionville
Pétionville	4	2	Belvil
Belvil	4	4	Sarthe

Differentiation

Some students may benefit through the use of calculators during this task. Others might need a number line to help in determining which affected areas have greater than 10,000 cholera outbreaks.

If students finish early, have them calculate the horizontal and vertical distances between affected areas that have not been previously calculated.

Technology Connection

Students could create a spreadsheet to track and graph the data.

Choices for Students

Following the introduction, offer students the opportunity to address a different area of Haiti other than the affected areas around Port-au-Prince.

Meaningful Context

Cholera, a potentially fatal illness, is spread through contaminated water. Cholera was already present in Haiti at the time of the January 2010 earthquake, and became widespread after the quake. Organizations are trying to mitigate the water crisis by installing chlorination and filtration systems, but it's difficult to determine which areas to target since the entire country is affected. Haitians often walk to their water sources because there is no public water sanitization system.

Recommended Resources

- Absolute Value
  www.walch.com/rr/CCTTG6AbsoluteValue
  This site provides a lesson on the concept of absolute value, followed by an interactive quiz that offers immediate feedback for each question.
- Distance Between Two Points
  www.walch.com/rr/CCTTG6TwoPoints
  Students can drag two points around a coordinate plane, then watch as the virtual graph paper calculates the distance between the points.
- International Action
  www.walch.com/rr/CCTTG6HaitiWater
  Students can learn more about the water crisis in Haiti and what one organization is doing to create a solution.
- Locate the Aliens
  www.walch.com/rr/CCTTG6LocateAliens
  To win this interactive game, students must correctly enter the coordinates of an alien displayed on the coordinate plane.

Water Crisis in Haiti

Part 1

On January 12, 2010, an earthquake devastated the country of Haiti. Hundreds of thousands lost their homes, their livelihoods, and access to fresh drinking water. Victims were forced to use water contaminated by human waste and trash for drinking, cooking, and washing. This led to outbreaks of cholera, a potentially fatal illness.

Since then, agencies across the globe have been working to help stop the spread of cholera. One strategy to limit the number of cholera outbreaks is to install chlorination or filtration systems to clean the water. You work for an organization that helps to provide clean water in Haiti. Your job is to determine which areas have been hit the hardest by creating a graph of outbreak data. The following table lists affected areas, their coordinates, and the numbers of reported cases of cholera.

Affected area	Coordinates	Approximate reported cholera cases
Belvil	(4, 3)	4,500
Calffebe	(4, –3)	6,600
Carrefour	(–5, 4)	10,700
Chafa	(–5, –1)	4,000
Cité Soleil	(–1, 7)	11,100
Delmas	(–1, 4)	5,600
Henry	(–6, –3)	3,700
Kenscoff	(2, –3)	10,300
Pétionville	(0, 1)	3,000
Port-au-Prince	(–2, 4)	12,000

Adapted from www.walch.com/CCTTG6HaitiCholera

NAME:

6.NS.8 Task • The Number System
Water Crisis in Haiti

1. Plot and label the affected areas on the coordinate plane of the map that follows.

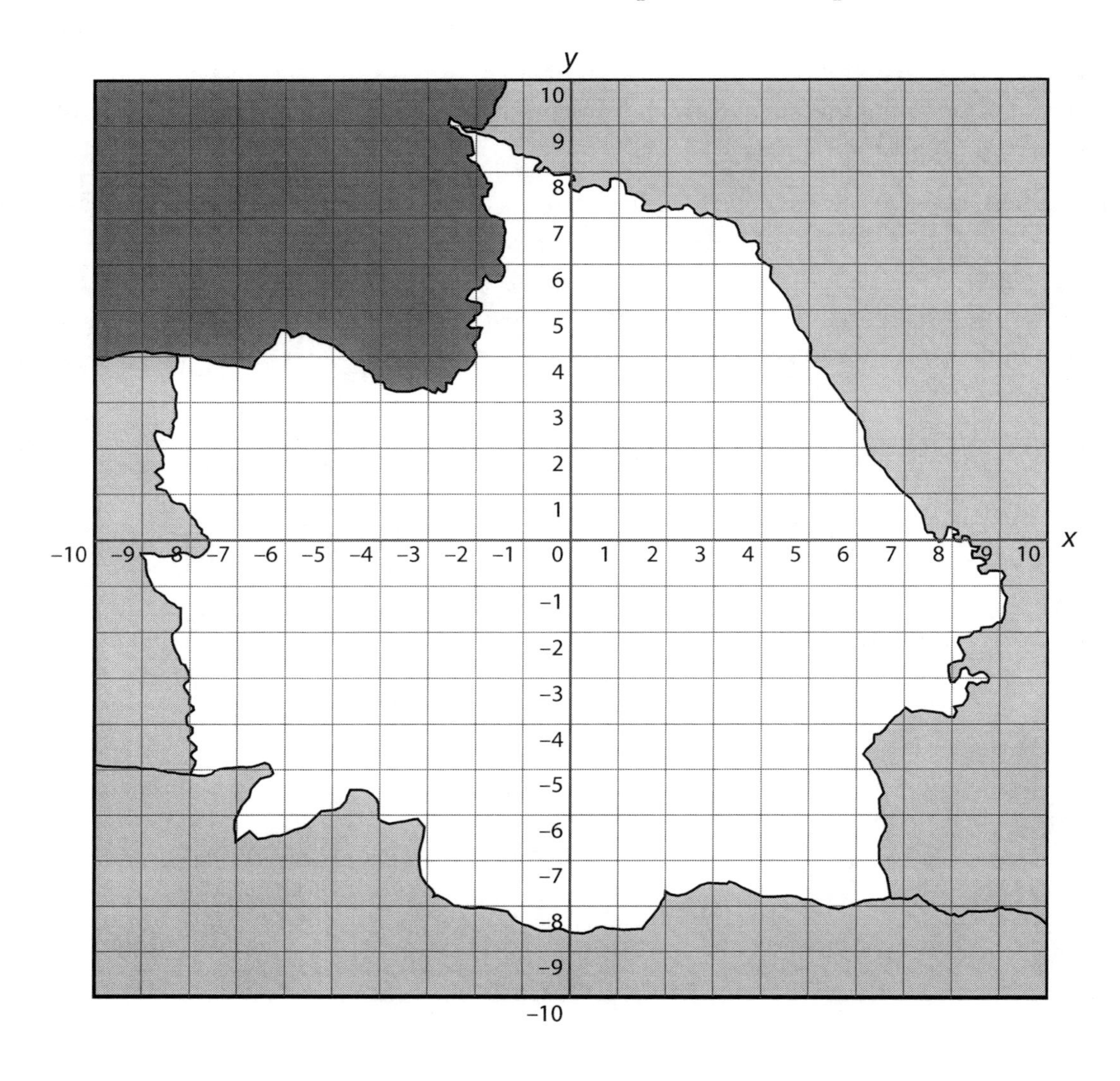

2. Mark any affected areas with more than 10,000 reported cholera cases in red.

3. Looking at your graph, which quadrant should organizations target first for chlorination and filtration systems? Explain your reasoning.

Part 2

You need to present your case for installing chlorination systems to your organization's board of directors. You must figure out the best order in which to install the systems in each affected area.

4. Explain how to use absolute value to calculate the distance between Henry and Calffebe. What is the distance between these two places in map units? ____________________

5. The airport is located in Sarthe, which has coordinates of (0, 7). Plot and label this place on your map.

6. Which affected area on your map is closest to the airport? ____________________

   Give the coordinates. ____________________

   Using the units on the map, how far away is the airport from this affected area?

   ____

   How did you calculate this distance?

7. You want to travel to the center of Port-au-Prince since this area has the most cases of reported cholera outbreaks. To get there, you travel through Delmas. What is the distance in map units to Delmas from the place you listed for question 6? ____________________

   How did you calculate that distance?

continued

8. Carrefour is the next affected area where you need to install chlorination and filtration systems. How far is Carrefour from Port-au-Prince in map units? ____________________

   Explain your answer.

9. Which affected area would you choose to help next? Explain your reasoning using horizontal distance, vertical distance, and absolute value.

10. Create a plan to visit the remaining affected areas. Write the order in which you would visit them and give the horizontal and vertical distances from each affected area to the next one. End at the airport. Use the table below to help organize your plan.

Originating area	Horizontal distance (x-axis)	Vertical distance (y-axis)	Destination

How Hot Is It?

Common Core State Standard

Apply and extend previous understandings of arithmetic to algebraic expressions.

6.EE.2c. Evaluate expressions at specific values of their variables. Include expressions that arise from formulas used in real-world problems. . . .

Task Overview

Background

One of the first steps in working with variables is to simplify algebraic expressions. Students often have difficulty extending their previous understanding of arithmetic to working with algebraic expressions. This difficulty can be alleviated by using real-world examples involving algebraic expressions with which students are familiar.

The task also provides practice with:

- order of operations
- grouping symbols
- operations with signed numbers
- creating graphs
- reading data tables

Implementation Suggestions

To break the task down into more manageable chunks, assign subsets of calculations to pairs or small groups of students when converting Celsius temperatures to Fahrenheit and vice versa. Have the pairs/groups post their results so that all the data is available for analysis.

Use of calculators is encouraged to minimize the amount of time students spend performing the calculations.

Introduction

Introduce the task by asking students what different ways there are to measure temperature. Ask if students have traveled to places where the Celsius scale is used to measure temperature. Question if students had difficulty trying to convert a Celsius temperature to a Fahrenheit temperature. Ask: Has anyone planned a trip or vacation to a place where temperatures are typically given using the Celsius scale? Did you have to convert the temperatures to Fahrenheit in order to know what type of clothing you should pack?

Monitoring/Facilitating the Task

Ask questions and prompt student thinking so that they:

- Recognize the mathematical operations they are using. Make sure that students articulate where and when they are using addition, subtraction, multiplication, and division during each calculation.
- Describe their arithmetic and explain their responses. Ask them to explain verbally how they determined their answers. Prompt for and encourage the use of proper mathematical terms.
- Realize that variables are used when different values can be used in an algebraic expression and that variables can be letters other than "x". Sometimes letters are used as variables in a meaningful way (for example, "F" is used to represent temperature Fahrenheit and "C" is used to represent temperature Celsius).
- Convert between two different temperature scales and round to the nearest degree accurately. Evaluate expressions using both positive and negative numbers.
- Multiply by a negative number and then add a negative number to a positive number (in problem 1). Use signed numbers in converting the temperatures (in problem 2).
- Correctly use the order of operations, including the use of grouping symbols.
- Use addition, subtraction, multiplication, division, and grouping symbols when converting from degrees Fahrenheit to degrees Celsius.
- Include multiple observations about the data in their responses to problems 9 and 10. Encourage students to use comparative terminology (greater than, less than, etc.) in their answers.
- Contribute to a discussion about how to determine an appropriate scale and how to label the axes. Use color or different line styles to distinguish between the temperatures for each location.

Debriefing the Task

- Ask students to share their data tables and graphs.
- Encourage students to explain how they used the formulas to convert the temperatures.
- Discuss the difference between an expression and an equation, noting that the conversion formula is an equation because it has an equal sign and is solved, rather than being evaluated.
- Encourage students to share where they had difficulty and how they resolved their questions.
- Encourage students to explain how they determined how to label their graphs (x- and y-axes) and how they determined the proper scale.
- Ask students if they had trouble graphing the data that included the negative values. Encourage students to explain how they adjusted their graphs to show the negative values.
- Encourage students to share their own experiences with different temperature scales. Examples could include vacationing abroad, or situations in which a family member might be overseas for military service or other work.
- Assess understanding by expanding the parameters used in the task. Encourage students to consider other real-world examples where algebraic expressions are evaluated. Ask students to explore how this task can relate to their life outside of the math classroom.

Answer Key

1. Moscow's January average temperature in degrees Fahrenheit is 18.
2. 19°F for February, 30°F for March, 43°F for April, 55°F for May, 63°F for June, 64°F for July, 61°F for August, 52°F for September, 41°F for October, 28°F for November, and 21°F for December
3. **Average Monthly Temperatures**

	Moscow (°F)	Iceland (°F)	Saudi Arabia (°F)	Honolulu (°F)
January	18°	32°	57°	72°
February	19°	33°	61°	72°
March	30°	36°	70°	73°
April	43°	39°	77°	74°
May	55°	45°	88°	76°
June	63°	50°	93°	78°
July	64°	55°	95°	79°
August	61°	53°	93°	80°
September	52°	48°	88°	79°
October	41°	42°	79°	78°
November	28°	37°	70°	76°
December	21°	32°	59°	73°

4. The highest temperature in degrees Fahrenheit occurs in Saudi Arabia in July, which has an average monthly temperature of 95°F. The lowest temperature in degrees Fahrenheit occurs in Moscow in January, which has an average monthly temperature of 18°F.

5.

Average Monthly Temperatures in Degrees Fahrenheit

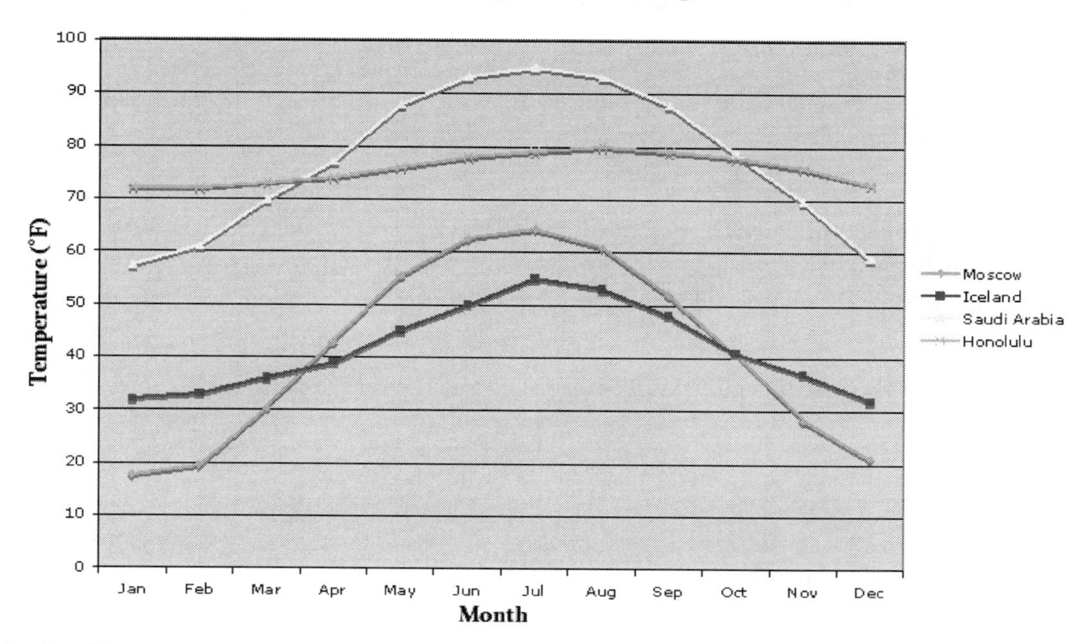

6. Iceland's average temperature in January in degrees Celsius is 0.
7. 1°C for February, 2°C for March, 4°C for April, 7°C for May, 10°C for June, 13°C for July, 12°C for August, 9°C for September, 6°C for October, 3°C for November, and 0°C for December

8.

Average Monthly Temperatures

	Moscow (°C)	Iceland (°C)	Saudi Arabia (°C)	Honolulu (°C)
January	–8°	0°	14°	22°
February	–7°	1°	16°	22°
March	–1°	2°	21°	23°
April	6°	4°	25°	23°
May	13°	7°	31°	24°
June	17°	10°	34°	26°
July	18°	13°	35°	26°
August	16°	12°	34°	27°
September	11°	9°	31°	26°
October	5°	6°	26°	26°
November	–2°	3°	21°	24°
December	–6°	0°	15°	23°

9. The highest temperature in degrees Celsius occurs in Saudi Arabia in July, which has an average monthly temperature of 35°C. The lowest temperature in degrees Celsius occurs in Moscow in January, which has an average monthly temperature of –8°C.

10. The times and places where the highest and lowest temperatures occur are the same regardless of the scale used. The scale is relative.

11. Sample answer: The graphs look identical except for the title and the scale on the y-axis. (Slight differences in the graphs are the result of rounding actual temperatures). Moscow and Iceland have the lower temperatures in both graphs and Saudi Arabia and Honolulu have the higher temperatures in both graphs. The shapes of the graphs are the same, with lower temperatures in January, February, and December and higher temperatures in June, July, and August.

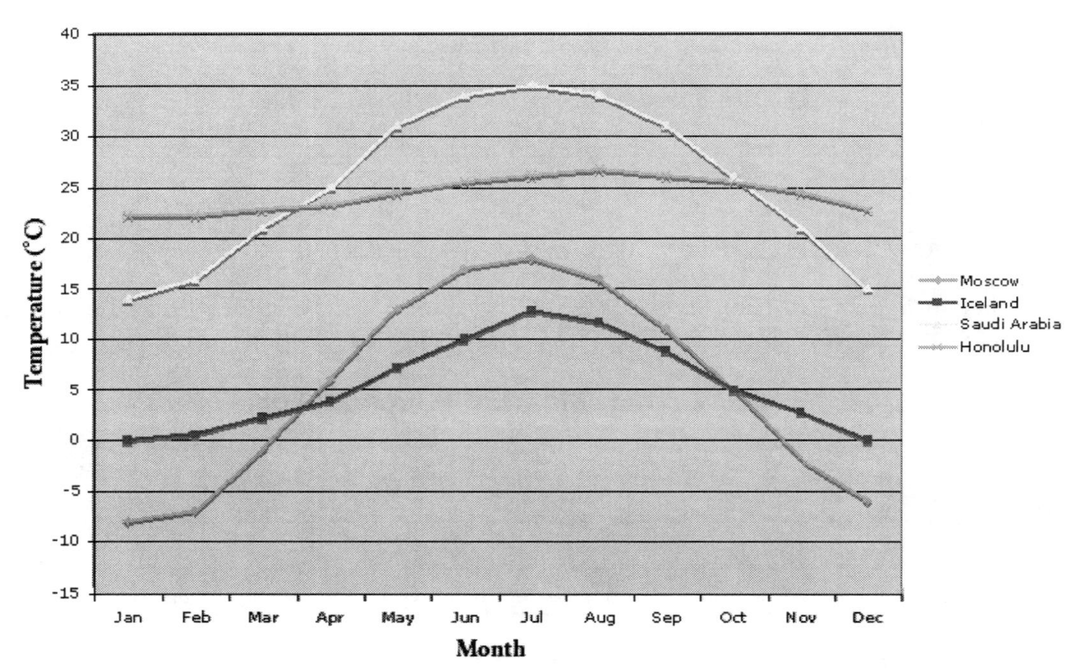

12. Sample answer: The formulas are similar since they both contain a fraction and the number 32. In one formula the fraction is 9/5, and in the other it is 5/9; the numbers are the same but are inverted. In one formula you need to add the 32 and in the other you need to subtract the 32. If you start with a temperature in °C, then convert it to °F, and finally convert it back to °C, you will get the same number you started with. These formulas "undo" each other.

Differentiation

Most students will benefit from the use of calculators. Alternatively, students could do the calculations by hand, but only compare two of the locations—one that has temperatures given in °C and the other in °F. Students could also use a graphing software program to create their graphs.

Technology Connection

Students could create a spreadsheet to record the temperatures in Fahrenheit and in Celsius.

Choices for Students

Following the introduction, offer students the opportunity to use temperatures from different places. Also offer students the option of creating a graph using graph paper and pencil, or by using technology.

Meaningful Context

This task may be expanded by having students research various places around the world to compare and contrast temperature data. Possible questions include:

- Where in the world is the highest temperature? Where is the lowest temperature?
- Where is the temperature difference the greatest? The lowest?
- Does anyone have a family member who lives in a location that typically uses the Celsius scale for temperatures? Where is this person? Find the average monthly temperature in degrees Fahrenheit for that location.

Recommended Resources

- edHelper.com
  www.walch.com/rr/CCTTG6EdHelper
  This Web site provides worksheets with extra problems for students who need additional practice with evaluating expressions. It also contains word problems, lesson plans, and activity ideas.
- Kuta Software
  www.walch.com/rr/CCTTG6KutaWorksheets
  Teachers can create worksheets with extra practice problems for students who need additional practice with evaluating expressions.
- Math.com
  www.walch.com/rr/CCTTG6MathResources
  This Web site includes teacher resources for a variety of topics.
- National Geographic Kids Atlases: Resources
  www.walch.com/rr/CCTTG6KidsAtlases
  This Web site contains resources for all sorts of subjects pertaining to the world community. There is a section where students can research average temperatures from around the world.

- Weather Explained: Record-Setting Weather
  www.walch.com/rr/CCTTG6RecordWeather
  This Web site contains information about the highest and lowest temperatures from around the world.
- Weather Wiz Kids: Temperature
  www.walch.com/rr/CCTTG6TemperatureConversion
  This Web site contains student-friendly information about weather. This particular page contains a temperature conversion tool.

6.EE.2(c) Task • Expressions and Equations
How Hot Is It?

Part 1

The table below shows the average monthly temperatures for four different places around the world.

Average Monthly Temperatures

	Moscow (°C)	Iceland (°F)	Saudi Arabia (°C)	Honolulu (°F)
January	–8°	32°	14°	72°
February	–7°	33°	16°	72°
March	–1°	36°	21°	73°
April	6°	39°	25°	74°
May	13°	45°	31°	76°
June	17°	50°	34°	78°
July	18°	55°	35°	79°
August	16°	53°	34°	80°
September	11°	48°	31°	79°
October	5°	42°	26°	78°
November	–2°	37°	21°	76°
December	–6°	32°	15°	73°

Notice that Moscow and Saudi Arabia have temperatures given in °C (degrees Celsius) while Iceland and Honolulu have temperatures given in °F (degrees Fahrenheit). In order to compare the temperatures in all four places, you will need to use the same temperature scale. Convert all the Celsius temperatures to Fahrenheit and graph the results.

- The equation to convert °C to °F is $F = \frac{9}{5} \times C + 32$, where *F* is the temperature in °F and *C* is the temperature in °C.
- The equation to convert °F to °C is $C = \frac{5}{9} \times (F - 32)$, where *F* is the temperature in °F and *C* is the temperature in °C.

NAME:

6.EE.2(c) Task • Expressions and Equations
How Hot Is It?

1. Convert the average temperature in Moscow in January from degrees Celsius to degrees Fahrenheit using the equation $F=\frac{9}{5}\times C+32$. Show your work. Round to the nearest degree. Add your answer to the table at the bottom of the page.

2. Continue to convert the rest of the average temperatures for Moscow from degrees Celsius to degrees Fahrenheit. Round to the nearest degree. Add your answers to the table below. Be prepared to explain how you got your answers.

3. Convert the average temperatures for Saudi Arabia from degrees Celsius to degrees Fahrenheit in the same manner. Round to the nearest degree. Complete the table below.

Average Monthly Temperatures

	Moscow (°F)	Iceland (°F)	Saudi Arabia (°F)	Honolulu (°F)
January		32°		72°
February		33°		72°
March		36°		73°
April		39°		74°
May		45°		76°
June		50°		78°
July		55°		79°
August		53°		80°
September		48°		79°
October		42°		78°
November		37°		76°
December		32°		73°

6.EE.2(c) Task • Expressions and Equations
How Hot Is It?

4. Which place has the highest temperature in °F? Which place has the lowest temperature in °F? Explain.

5. Create a graph showing the average monthly temperatures of all four places on the same graph and using temperatures in degrees Fahrenheit.

Part 2

You would like to share this information with a friend from Canada who is much more familiar with temperatures given in degrees Celsius. Convert all the Fahrenheit temperatures to Celsius and graph the results. Compare this graph and the formula you used to the graph and formula from Part 1.

6. Convert the average monthly temperature in Iceland in January from degrees Fahrenheit to degrees Celsius using the equation $C=\frac{5}{9}\times(F-32)$. Round to the nearest degree. Show your work. Add your answer to the table below.

7. Continue to convert the rest of the average temperatures for Iceland from degrees Fahrenheit to degrees Celsius. Round to the nearest degree. Add your answers to the table below. Be prepared to explain how you got your answers.

8. Convert the average temperatures for Honolulu from degrees Fahrenheit to degrees Celsius in the same manner. Complete the table below.

Average Monthly Temperatures

	Moscow (°C)	Iceland (°C)	Saudi Arabia (°C)	Honolulu (°C)
January	–8°		14°	
February	–7°		16°	
March	–1°		21°	
April	6°		25°	
May	13°		31°	
June	17°		34°	
July	18°		35°	
August	16°		34°	
September	11°		31°	
October	5°		26°	
November	–2°		21°	
December	–6°		15°	

9. Which place has the highest temperature in °C? Which place has the lowest temperature in °C? Explain.

10. How does your answer to problem 9 compare with your answer to problem 4? Fully explain.

11. Create a graph showing the average monthly temperatures of all four places on the same graph and using temperatures in degrees Celsius. How does this graph compare with the graph you created in problem 5?

12. Compare the two formulas. What is similar? What is different? What happens if you start with a temperature in °C, then convert it to °F, and finally convert it back to °C? Explain.

6.EE.4 Task • Expressions and Equations

Eat Local

Instruction

Common Core State Standard

Apply and extend previous understandings of arithmetic to algebraic expressions.

6.EE.4. Identify when two expressions are equivalent (i.e., when two expressions name the same number regardless of which value is substituted into them). For example, the expressions $y + y + y$ and $3y$ are equivalent because they name the same number regardless of which number y stands for.

Task Overview

Background

Students often struggle when working with variables. They have difficulty with the abstract ideas associated with a letter or symbol representing a numerical value.

Some students will have difficulty identifying when two expressions are equivalent. This task is designed to help students effectively work with variables in expressions in the context of a real-world problem, including understanding that expressions can look different and still have the same value. This can be connected to their prior knowledge that 5 + 5 + 5 is the same as 3 • 5. Connecting algebraic expressions with real-world fencing lengths will also help students to understand equivalent expressions. In this case, they will be looking for combinations of $1x$, $2x$, and $3x$ that are equivalent to $8x$.

The task also provides practice with:

- addition and multiplication
- the order of operations
- finding perimeter
- calculating area
- drawing diagrams

Implementation Suggestions

Students may work individually, in pairs, or in small groups to complete one or both parts of the task.

Alternatively, students may meet in groups to share their results and reflect after individually completing the task, before a class discussion.

It may be helpful to have concrete materials available for students to use to model the fencing lengths. Unifix cubes would be helpful in modeling x, $2x$, and $3x$. Graph paper is also helpful for visualizing unit lengths and area, especially the idea of square units.

6.EE.4 Task • Expressions and Equations
Eat Local

Introduction

Introduce the task by asking students whether they have ever helped someone put up a fence, build a patio, or put in a garden. Have they helped calculate the area of the space? Have they helped figure out how much material (fencing) to purchase? Briefly discuss these experiences and note students' apparent understanding of area.

You might also quickly review the idea of equivalent expressions. Ask students if they recall exercises where they listed all the ways to make 100 (for example): $10 \bullet 10$, $10 + 90$, $20 + 20 + 20 + 20 + 20$, etc. Explain that the same is true for expressions with variables, so $3x = x + x + x$ or $2x + x$, and so forth, and that this is what they'll be exploring in this task.

Monitoring/Facilitating the Task

Ask questions and prompt student thinking so that they:

- Make sense of what the task is asking for, which includes seeing the fencing material in units that can be represented algebraically, with the smallest length becoming the "base unit" and the others described relative to that unit.
- Recognize the mathematical operations they are using. Make sure that students articulate where and when they are using addition and multiplication during each calculation.
- Describe their arithmetic and explain/defend their responses. Ask them to explain verbally how they determined their answers. Prompt for and encourage the use of proper mathematical terms.
- Realize that variables are used when different values can be substituted in an algebraic expression, that the variables can be letters other than "x," and that different variables can be used in one expression to represent different things of unknown quantity.
- Simplify and expand expressions with variables.
- Recognize that a number beside a letter without an operation symbol (e.g., $3m$) indicates multiplication.
- Correctly use the order of operations. The calculations in this task require the use of addition and multiplication.
- Sketch accurate diagrams.

Debriefing the Task

Encourage students to:

- Explain how the expressions can look different but be equivalent.
- Use appropriate mathematical terminology.
- Share where they had difficulty and how they resolved their questions.
- Explain how they sketched their diagrams.
- Describe their own experiences with planting a garden, helping to fence in a yard, or staking out an animal enclosure. What steps were taken to make sure the project was completed correctly?
- Consider other real-world examples where equivalent algebraic expressions are used and evaluated. An example of this is the expression to convert temperature from Celsius to Fahrenheit (or the reverse). Another instance occurs in forensic science, in which the height of a person can be deduced from the length of one of their bones. Both these examples extend the fundamental concept introduced in the task, but simpler examples can also be used.
- Explore how this task can relate to their lives outside of the math classroom.

Assess understanding by expanding the parameters used in the task. For example, how would the problem be different if the length and width of the garden space were changed?

Answer Key

1. Let x represent the length of the smallest piece of fencing. Note that any letter could be used here as the variable. The other two pieces would be $2x$ and $3x$.

   x $2x$ $3x$

2. The length of the side of the wing is $8x$.

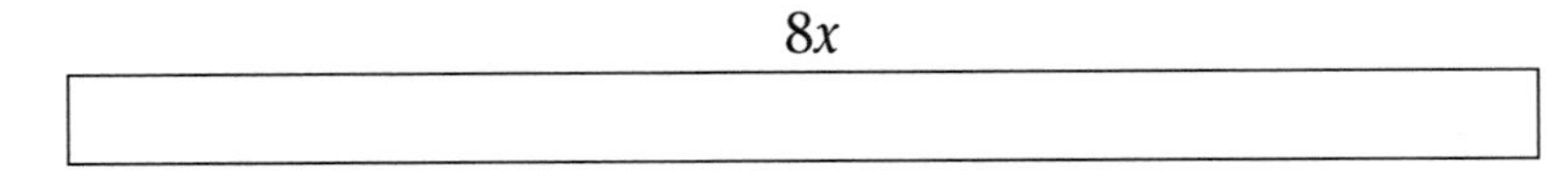

3. Answers will vary. Sample answer: $3x + 3x + 2x$

   $3x$ $3x$ $2x$

4. Answers will vary. Sample answer: $x + x + x + 2x + 3x$

x x x $2x$ $3x$

5. Answers will vary. Sample answer: Both expressions represent the same length. If you simplify the expressions, they are equivalent to $8x$. They are different representations of the same length.

6. The width of the garden space is $4x$.

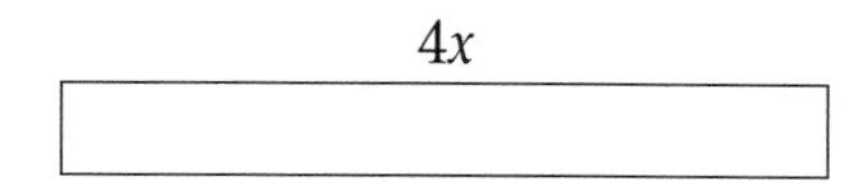

7. Answers will vary. Sample answer: $x + 3x$ and $2x + 2x$.

x $3x$ $2x$ $2x$

8. Answers may vary but should include a drawing that is twice as long as it is wide and that is labeled $8x$ and $4x$. For example:

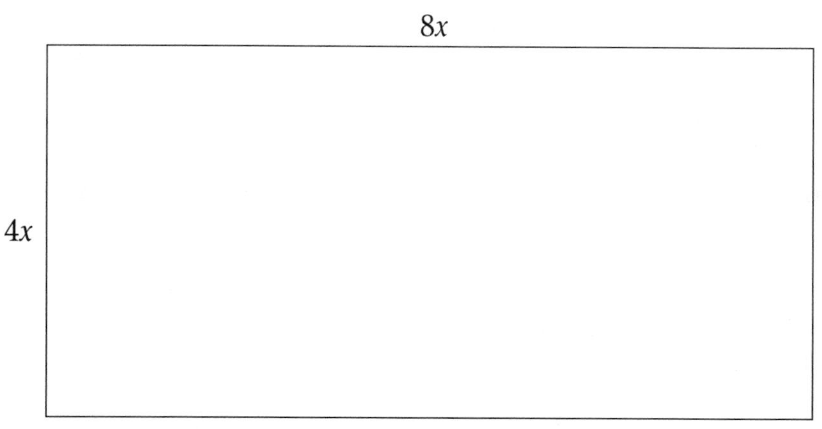

9. $8x + 8x + 4x + 4x = 24x$

10. The total fencing needed is 24 • 3 = 72 feet. The garden space's area is 8(3) • 4(3) = 288 square feet.

11. Student responses will vary. Check for completion and accuracy.

Differentiation

Some students may benefit through the use of calculators during this task. Some may choose to, or need to, use concrete materials to model the situations. Students could use a drawing software program to create their diagram, or use what they learn to create their own community garden.

Technology Connection

Students could use a drawing program to create the drawing of the fenced-in garden space.

Choices for Students

Following the introduction, offer students the opportunity to plan their own garden space, using different lengths of fencing. They might also designate a different situation, such as a different enclosure that is more meaningful or familiar to them (fencing for a pet, etc.).

Meaningful Context

This task may be expanded by having students calculate how much total fencing is needed to fence in the garden space, remembering that one side of the area is the side of the wing of the school. The students could then research the costs of different types of fencing and calculate the cost of the materials for the project. Again, they could apply these procedures and concepts to a more familiar or meaningful context, such as fencing for a pet.

Recommended Resources

- American Community Garden Association
  www.walch.com/rr/CCTTG6CommunityGarden
  Students can learn about the community garden movement, and about how to find a local community garden or start one if there isn't one nearby.
- Eat Local
  www.walch.com/rr/CCTTG6EatLocal
  This site offers information on ways to add more local produce to your diet.
- Home Depot
  www.walch.com/rr/CCTTG6HomeDepot
  Students can research the cost of fence material at this home store's Web site.
- Lowe's
  www.walch.com/rr/CCTTG6Lowes
  Students can research the cost of fence material at this home store's Web site.
- NCTM Illuminations—Pan Balance for Expressions
  www.walch.com/rr/CCTTG6PanBalance
  This interactive resource allows teachers or students to enter two expressions and to see their relationship in terms of a pan balance that is either evenly balanced or tipped to one side.

6.EE.4 Task • Expressions and Equations
Eat Local

Part 1

You're very enthusiastic about the "eat local" movement, but in your city there are very few places to get fresh produce. Using information from the American Community Garden Association, you're helping to build a community garden on the grounds of your school. You decide to fence a garden space that is as long as one wing of the school and half as wide. What is the total amount of fencing that you need, and what will the area of the garden space be?

The fence material you will use is available in a variety of sizes. You do not know the size of the smallest piece of fencing, but you do know that there are additional pieces available that are 2 times and 3 times as long as the small piece. Since you have to build a fence and don't know the specific lengths of the pieces of fence, you need to figure out how to think about the various lengths and how to plan the garden space.

You also decide that it would be helpful to sketch drawings of your plans before you begin work.

1. How can you use algebraic notation to represent the length of each of the 3 available sizes of fencing? How can you draw what a piece of each size looks like?

2. If the length of the school wing is 8 times longer than the smallest piece of fencing, how can you represent the length of the side of the wing in terms of the length of the smallest piece of fencing? Draw a diagram to show what the length of the wing of the school looks like.

continued

6.EE.4 Task • Expressions and Equations
Eat Local

3. Using more than one size of the available fencing, how could you create one side of the garden space that is the same length as the wing of the school? How can you use algebraic notation to represent the length of the garden space? Draw a diagram to show what the length of the garden space looks like using the available fencing.

4. Is there a different way to create the length of the garden space also using more than one size of fencing? How could you use algebraic notation to represent this length of the garden space? Draw a different diagram showing what the length of the garden space looks like using the available fencing.

5. Look at your answers for problems 3 and 4. How are they the same? How are they different?

continued

6.EE.4 Task • Expressions and Equations
Eat Local

Part 2

Now you need to plan the width of the garden space. You have decided to make the garden space half as wide as it is long. Since you still do not know the actual length of the fencing, you need to figure out how to put the various pieces of fence together to make the garden space.

6. How could you represent the width of the garden space using algebraic notation and in terms of the smallest available size of fencing? Draw what the width of the garden space looks like.

7. Can you think of two different ways to create the width of the garden space using more than one size of fencing? How could you use algebraic notation to represent these widths of the garden space?

8. How can you show what the entire garden space looks like?

continued

6.EE.4 Task • Expressions and Equations
Eat Local

9. How can you calculate the total amount of fencing you need in terms of the smallest size of fencing?

10. You find out that the smallest piece of fencing is 3 feet long. How could you determine the total amount of fencing you need? How could you determine the area of the garden space?

11. Summarize your plan for building a garden space. Use algebraic equations to represent the length and width of the garden space and to show the area.

Fund-raising

Common Core State Standard

Reason about and solve one-variable equations and inequalities.

6.EE.7. Solve real-world and mathematical problems by writing and solving equations of the form $x + p = q$ and $px = q$ for cases in which p, q, and x are all nonnegative rational numbers.

Task Overview

Background

Solving simple equations is one of the fundamental learning goals in algebra. In order to complete this task, students should have prior experience with algebraic expressions and solving simple equations.

This task is designed to give students practice with solving simple equations in the real-world context of selling tickets for a fund-raiser. Students will write equations from a verbal representation of the problem statement and then solve the equation. In Part 1, the equations will be solved using subtraction; in Part 2, the equations will be solved using division.

The task also provides practice with:

- adding and subtracting positive and negative numbers
- writing an equation from a verbal expression

Implementation Suggestions

Students may work individually, in pairs, or in small groups to complete one or both parts of the task. Students may use one of the virtual manipulative tool sites provided in the Recommended Resources as they complete the task.

Introduction

Introduce the task by asking students if they have been involved in a fund-raising activity. What types of fund-raising activities were they? Has anyone helped keep track of the number of tickets sold or that need to be sold? How did they do that?

If necessary, have students recall the difference between an algebraic expression and an algebraic equation (an equation contains an equal sign). As appropriate, review how to write an algebraic equation given a verbal representation of a problem statement.

Determine whether students will work individually or in pairs and make the appropriate groupings.

Monitoring/Facilitating the Task

Ask questions and prompt student thinking so that they:

- Write algebraic equations to represent the problem statement and then solve the equation. Some students will be able to answer the questions without using a formal algebraic equation.
- Think about the variable they choose to use in the equation. Does it have to be x or could it be something else?
- Carefully read each question and make the appropriate changes to their equations.
- Think about other examples in which these types of algebraic equations are used to solve problems. Ask students to be specific in their examples.

Debriefing the Task

- Upon completion of the task, students should share their work.
- Ask students to discuss how they formulated their equations, looking closely at how a real-life situation was represented algebraically.
- Discuss the choice of variables and why it is sometimes useful to use a variable other than x. (Meaningful variables can help you remember what the variable represents—for example, n can be used to represent the number of something, and t can be used to represent time.)
- Ask students to explain the steps they used to solve their equations using appropriate mathematical terminology (for example, subtract from both sides of the equation or divide both sides of the equation by ...).
- Ask students to share any difficulties they had with this task. Encourage them to describe how they addressed these difficulties.
- Encourage students to share their other real-life examples in which using algebraic equations of this type would be appropriate. Ask them to describe why they picked these examples and what makes them similar mathematically.

Answer Key

1. $x + 175 = 600$
2. Subtract 175 from both sides of the equation: $x = 425$; 425 tickets remain to be sold.
3. $x + 395 = 600$
4. Subtract 395 from both sides of the equation: $x = 205$; 205 tickets remain to be sold.
5. $x + 175 = 500$ and $x + 395 = 500$; after the first day 325 tickets remain to be sold, and after the third day 105 tickets remain.
6. $5y = 175$
7. Divide both sides of the equation by 5; each student sold 35 tickets.
8. $5y = 395$
9. Divide both sides of the equation by 5; each student sold 79 tickets.
10. Answers will vary. Sample answers:

    Jake and Kate have a total of 25 marbles. Jake has 13 marbles. How many marbles does Kate have? $n + 13 = 25$; Kate has 12 marbles.

    Four students are playing a card game. 104 cards need to be evenly distributed among the players. How many cards should each player receive? $4c = 104$; each player should receive 26 cards.

Differentiation

Some students may benefit from the use of calculators during this task. Some students may find it helpful to use algebra tiles.

Students who complete the task early could research prices of MP3 players, determine an amount of money to be raised, and then determine a ticket price for the raffle to meet that goal.

Technology Connection

Students could use one of the virtual manipulative tools prior to or while completing the task.

Choices for Students

Following the introduction, offer students the opportunity to create their own example of a fund-raising idea that could be represented by one of these types of algebraic equations.

Meaningful Context

This task uses a real-world example of fund-raising. Most students will be involved in some aspect of fund-raising during their school years or beyond. This task shows how formal algebraic equations can be used to represent real-world problems and help solve the problems.

Recommended Resources

- Algebra Balance Scales
  www.walch.com/rr/CCTTG6AlgebraBalance
  This virtual manipulative allows students to solve simple linear equations through the use of a balance beam.
- Algebra Tiles
  www.walch.com/rr/CCTTG6AlgebraTiles
  Using tiles to represent variables and constants, this virtual manipulative shows how to represent and solve an algebra problem.
- Equation Balance
  www.walch.com/rr/CCTTG6EquationBalance
  This applet gives students practice with solving simple algebra equations using an equation balance.

6.EE.7 Task • Expressions and Equations
Fund-raising

Part 1

The local Boys and Girls Club is raffling off an MP3 player to raise money for its programs. There were 600 tickets printed. You are in charge of keeping track of how many tickets have been sold and, likewise, how many remain to be sold. Use algebraic equations to help you perform your calculations.

1. After the first day of sales, 175 tickets have been sold. How could you write an algebraic equation using addition that could be used to find how many tickets remain?

2. Use this equation to determine the number of tickets that remain to be sold. Explain your steps.

3. After the third day of sales, the number of tickets sold totals 395. How could you write a new equation using addition that could be used to find how many tickets remain?

4. Use this new equation to determine the number of tickets that remain after the third day of sales. Explain your steps.

5. Your group decides that you will sell only 500 raffle tickets and save the rest of the tickets for another raffle. How would you revise your equations to reflect this change? Use the new equations to determine the number of tickets that remain to be sold after the first day of sales, and then after the third day of sales.

continued

Fund-raising

Part 2

There are 5 students selling tickets in your group. You want to determine how many tickets each student has sold. They tell you that they all sold an equal number of tickets for each day of sales.

6. Recall that after the first day, 175 tickets had been sold. How could you write an algebraic equation using multiplication that could be used to determine how many tickets each student sold?

7. Use this equation to determine how many tickets each student sold. Explain your steps.

8. Recall that after the third day, a total of 395 tickets had been sold. How could you write an algebraic equation using multiplication that could be used to determine how many tickets each student sold if they each sold the same amount?

9. Use this equation to determine how many tickets each student sold. Explain your steps.

10. Imagine that you are dividing marbles or dealing cards. Describe a situation for which you could use these kinds of algebraic equations to solve a problem. Write examples for both types of equations you used. Be specific and explain fully.

How Much Should That Specialty Pizza Cost?

Common Core State Standard

Represent and analyze quantitative relationships between dependent and independent variables.

6.EE.9.1. Use variables to represent two quantities in a real-world problem that change in relationship to one another; write an equation to express one quantity, thought of as the dependent variable, in terms of the other quantity, thought of as the independent variable. . . .

Task Overview

Background

This task allows students to think about independent and dependent variables in the real world. Students then represent the relationship between the two variables using an equation. Part 1 guides the students through the procedure for a small pizza, then asks them to write an equation for the medium and large pizzas independently. Part 2 asks students to consider the cost of specialty pizzas in relation to the "create your own" pizzas. Students must then determine if customers who order specialty pizzas are paying more than customers who create their own, and when it is cost effective to order a specialty pizza versus creating their own.

This task also provides practice with:

- writing equations to represent real-world situations
- comparing values
- calculating with decimals/money

Implementation Suggestions

Students can work individually or in groups for this task. This task should follow an introduction or review of independent and dependent variables.

Introduction

Ask students to think about dependent and independent variables in their lives. Have students suggest quantities that are dependent on other quantities. Examples may include cell phone bills and number of texts sent, or the final score of a baseball game and the total number of hits.

Monitoring/Facilitating the Task

Ask questions and prompt student thinking so that they:

- Notice that the responses for questions 1–7 use only two possible answers: the cost of pizza and the number of toppings.
- Realize that the cost of the pizza changes based on the number of toppings selected, and that this differs from the number of toppings changing based on the cost of the pizza.
- Select the correct variable, x or y, to represent their quantities.
- Justify their solutions in question 12 using the equations they created.

Debriefing the Task

Have students form small groups to discuss their results. Students should compare their equations, looking for similarities and differences. They should debate any inconsistencies in their solutions, using their equations to justify the answer they believe to be correct. Each group should decide on their group solutions to question 12 and write their answers on the board.

As a class, review the solutions to the worksheet, leaving the information on the board for the final discussion. Topics for discussion include:

- how students determined which quantity was dependent and which was independent
- how students represented the different information in an equation

Answer Key

1. the cost of the pizza and the number of toppings
2. the number of toppings
3. the cost of the pizza
4. cost of the pizza; number of toppings
5. number of toppings; cost of the pizza
6. the number of toppings; x
7. the cost of the pizza; y
8. cost of pizza = \$6.00 + number of toppings • \$0.75
9. $y = 6 + 0.75x$
10. $y = 8 + x$
11. $y = 11 + 1.25x$

12. Students should show evidence of having used their equations to arrive at the following conclusions for each type of pizza:

    I Love Meat: It's cheaper to create your own for all three sizes.

    I Love Vegetables: It's cheaper to order the specialty pizza for all three sizes.

    I Love Cheese: It costs the same to create your own small or medium as it would to order the specialty pizza, but the specialty large is cheaper than creating your own.

    I Love Simplicity: It's cheaper to create your own for the small and medium, but not the large.

    I Love Everything: It's cheaper to create your own for the small and medium, but not the large.

    Completed table:

Specialty pizza: I Love Meat			
	Small	Medium	Large
Specialty price	$12.00	$16.25	$19.50
"Create your own" price	$10.50	$14.00	$18.50
Specialty pizza: I Love Vegetables			
	Small	Medium	Large
Specialty price	$9.50	$13.75	$17.00
"Create your own" price	$12.75	$17.00	$22.25
Specialty pizza: I Love Cheese			
	Small	Medium	Large
Specialty price	$8.25	$11.00	$13.25
"Create your own" price	$8.25	$11.00	$14.75
Specialty pizza: I Love Simplicity			
	Small	Medium	Large
Specialty price	$7.00	$9.50	$11.50
"Create your own" price	$6.75	$9.00	$12.25
Specialty pizza: I Love Everything			
	Small	Medium	Large
Specialty price	$15.50	$19.00	$22.75
"Create your own" price	$13.50	$18.00	$23.50

Differentiation

You may consider assigning only one size or type of pizza to specific individuals in place of completing question 12 for all sizes and types. Students may also wish to design their favorite pizza and determine its cost if they created their own pizza with the available toppings.

Technology Connection

As a next step, to extend this task to the second part of the standard ("analyze the relationship between dependent and independent variables using graphs and tables..."), students can use graphing software to produce graphs and/or spreadsheet software to produce tables using the given information and their answers.

Choices for Students

Students may decide to use the menu from a local pizzeria to create the same comparisons. This information could then be used to create a "Best Pizza Deal at Joe's Pizza" pamphlet for circulation in the class, school, or community.

Meaningful Context

Independent and dependent variables are seen in many different contexts aside from those used in mathematics. Students can explore independent and dependent variables in a variety of contexts. For example, if global warming is the independent variable, have students compose a list of possible dependent variables that may be related to the independent variable. Have students research quantitative values to represent each variable.

Recommended Resources

- Independent vs. Dependent Variables
  www.walch.com/rr/CCTTG6IndependentVsDependent
  This animated video covers the basics of dependent and independent variables by comparing player performance in a soccer game.
- Internet Pizza Server Ordering Area
  www.walch.com/rr/CCTTG6PizzaEquations
  This site allows you to select traditional as well as unusual/impossible pizza toppings and "order" a pizza. The application then returns your pizza along with the cost of the pizza. Advanced students may enjoy the challenge of trying to figure out the equation used to represent the cost of the pizza.
- One Thing Depends on Another
  www.walch.com/rr/CCTTG6VariableVideo
  This music video spoof of an '80s pop song provides a short review of independent and dependent variables and includes domain, range, and function information.

NAME:

6.EE.9.1 Task • Expressions and Equations
How Much Should That Specialty Pizza Cost?

Part 1

Use what you know about independent and dependent variables along with the following information to answer the questions.

> A local pizza shop offers the option of creating your own pizza. Each pizza includes tomato sauce, mozzarella cheese, a thin crust, and additional toppings as requested. A small cheese pizza costs $6.00, and each additional topping is $0.75.

1. What are the two quantities that will change when deciding which pizza to order?

   ____

2. Which quantity represents the independent variable? ______________________

3. Which quantity represents the dependent variable?______________________

4. The ______________________ is dependent on the ______________________.

5. The ______________________ is independent of the ______________________.

6. The independent quantity, ______________________, is represented with the variable x OR y (circle one).

7. The dependent quantity, ______________________, is represented with the variable x OR y (circle one).

8. Write an equation to represent the cost of a small pizza in terms of the number of toppings ordered.

continued

How Much Should That Specialty Pizza Cost?

9. Substitute variables into the equation you wrote for problem 8.

The pizza shop also offers medium and large "create your own pizza" options. The medium pizza costs $8.00 plus $1.00 per topping. The large pizza costs $11.00 plus $1.25 per topping.

10. Use the same variables you chose in problem 9 to write an equation to represent the cost of a medium pizza in terms of the number of toppings ordered.

11. Use the same variables as in problem 9 to write an equation to represent the cost of a large pizza in terms of the number of toppings ordered.

continued

6.EE.9.1 Task • Expressions and Equations
How Much Should That Specialty Pizza Cost?

Part 2

The pizza shop also offers the following specialty pizzas. Each pizza includes tomato sauce, mozzarella cheese, a thin crust, and additional toppings as listed.

- **I Love Meat**
  pepperoni, ham, ground beef, bacon, sausage, chicken

  Small: $12.00 Medium: $16.25 Large: $19.50

- **I Love Vegetables**
  green peppers, tomatoes, onions, olives, mushrooms, spinach, jalapenos, broccoli, garlic

  Small: $9.50 Medium: $13.75 Large: $17.00

- **I Love Cheese**
  ricotta cheese, Parmesan cheese, Asiago cheese

  Small: $8.25 Medium: $11.00 Large: $13.25

- **I Love Simplicity**
  pepperoni

  Small: $7.00 Medium: $9.50 Large: $11.50

- **I Love Everything**
  pepperoni, ham, ground beef, bacon, sausage, chicken, green peppers, tomatoes, onions, olives

  Small: $15.50 Medium: $19.00 Large: $22.75

6.EE.9.1 Task • Expressions and Equations
How Much Should That Specialty Pizza Cost?

12. An ad in the pizza shop says you can "Save by ordering our specialty pizzas!" But since the name of the shop is Cheatza Pizza, you wonder if ordering a specialty pizza actually costs more than creating your own pizza with the same number of toppings. You want to know if customers are being cheated.

    Use the equations you created in Part 1 to find out whether it is cheaper to purchase the specialty pizza or to create your own for each specialty pizza combination. Fill in the "create your own" prices for each pizza in the table below. Is it ever cheaper to create your own pizza? Justify your response for each specialty pizza by using your equations.

Specialty pizza: I Love Meat			
	Small	Medium	Large
Specialty price	$12.00	$16.25	$19.50
"Create your own" price			
Specialty pizza: I Love Vegetables			
	Small	Medium	Large
Specialty price	$9.50	$13.75	$17.00
"Create your own" price			
Specialty pizza: I Love Cheese			
	Small	Medium	Large
Specialty price	$8.25	$11.00	$13.25
"Create your own" price			
Specialty pizza: I Love Simplicity			
	Small	Medium	Large
Specialty price	$7.00	$9.50	$11.50
"Create your own" price			
Specialty pizza: I Love Everything			
	Small	Medium	Large
Specialty price	$15.50	$19.00	$22.75
"Create your own" price			

Planning a Trip

Common Core State Standard

Represent and analyze quantitative relationships between dependent and independent variables.

6.EE.9.2. ... Analyze the relationship between the dependent and independent variables using graphs and tables, and relate these to the equation. For example, in a problem involving motion at constant speed, list and graph ordered pairs of distances and times, and write the equation $d = 65t$ to represent the relationship between distance and time.

Task Overview

Background

Understanding the relationship between variables is one of the building blocks for algebra. This task is designed to help students understand the relationship between independent and dependent variables. In order to complete this task, students should be able to create a table given data and then graph the data points. They should be able to determine how to label the axes and the scale that needs to be used.

This task helps students explore the relationship between independent and dependent variables in the context of a real-world example. Distance traveled versus time is a common example. Here speed is assumed to be constant. Many students will understand that the distance traveled depends upon the time spent travelling and that you can't change the distance without changing the time. You can change the time driving directly, by driving for a longer period of time. Distance can only be changed indirectly by changing the time driving. Therefore, time is the independent variable and distance is the dependent variable. Another example of this is average temperature and the month of the year. Clearly, the month of the year does not depend upon the average temperature. However, the average temperature does depend on the month of the year. Here the month would be the independent variable and the average temperature would be the dependent variable.

The task also provides practice with:

- creating a table given data
- plotting points and drawing a graph
- comparing graphs
- solving simple equations

Implementation Suggestions

Students may work individually, in pairs, or in small groups to complete one or both parts of the task.

Students will use one of the map sites to find the exact mileage to their destination and use that figure to complete the task.

Introduction

Introduce the task by asking students if they have ever been on a long car trip. Where did they go and how long did it take? How long did they travel each day? Has anyone helped plan a trip? How did they help?

Ask students, "How can you estimate the time that it will take to travel a certain distance? If you assume that speed is constant, how do you determine how far you will travel in a certain amount of time?"

If desired, you may want to review some of the procedures for setting up tables and graphing data before beginning the task.

You might begin the task by having students review the process of setting up a table of values and use that to create a graph. Make sure that students remember to use units (hours and miles) when labeling their table and graph.

As a class, review the process of creating a graph given data. Discuss how to label the axes and determine a scale.

Determine whether students will work individually or in pairs and make the appropriate groupings. Students who have selected the same destination should work in the same group.

Make certain that students have the materials they need to complete the task, including graph paper.

Monitoring/Facilitating the Task

Ask questions and prompt student thinking so that they:

- Will be able to select an interesting destination. Students could research places to visit in the United States. Alternatively, provide them with a list of suggestions, such as the Grand Canyon, Yosemite National Park, Disneyland, Cooperstown, or Niagara Falls.
- Accurately create the table of values.
- Correctly set up their graph with time on the x-axis. Make sure they understand why it is important to do this (the x-axis is used for the independent variable).
- Fully analyze the graph and describe it using correct mathematical terms (distance traveled increases as time increases, the line rises to the right, the slope is positive, etc.).

- Recognize that speed is a constant here. The two things that are changing are time and distance.
- Think about how to label the variables in order to make them more meaningful; for example, d could be used to represent distance and t to represent time.

Debriefing the Task

- Upon completion of the task, the students should share their results. Make sure to identify the destinations students chose and the estimated mileage they used.
- Encourage a discussion around why different destinations were chosen. What would be interesting about visiting that location?
- Ask students to share their graphs. Facilitate a discussion around how they determined the scale to use. Encourage students to share what is considered a completely labeled graph (a title, a legend to identify each graph, and axes labeled including units). Encourage a discussion about why it is important to label the graph and use units (to make the graphs understandable in the context of the task).
- Include a discussion around what the purpose of the task is. Is it to solve a mathematical problem or to develop skills that can be used in real life?
- Initiate a discussion about the factors that might make the formula $d = rt$ close, but not exact—traffic, bad weather, inconsistent speed, stopping for food or gas, etc. Consider the possibility of estimating the time that any or all of these might require and how that could be factored into trip plans.
- Encourage students to share their observations about their graphs. Specifically, students should notice that the slopes are both positive but one is greater than the other (the line is steeper). Encourage a discussion about how this relates to the equation generated.
- Ask students to explain the relationship between dependent and independent variables. Ask them to share their general statements.
- Promote discussion around other examples of dependent and independent variables. Other examples include season (independent) and foliage growth (dependent), time (independent) and mold growth (dependent), and amount of street garbage (independent) and raccoon population (dependent).
- Ask students to explore how this task can relate to their lives outside of the math classroom. For example, being able to analyze the relationship between independent and dependent variables can help understand cause and effect relationships in a scientific experiment.

Answer Key

Answers will vary depending upon the destination chosen by the students. Sample answers for a trip from Maryland to the Grand Canyon (about 1,900 miles) are given below to use as a guide.

1.

Time (hours)	Distance (miles)
1	55
3	110
5	275
10	550
20	1100
40	2200

2.

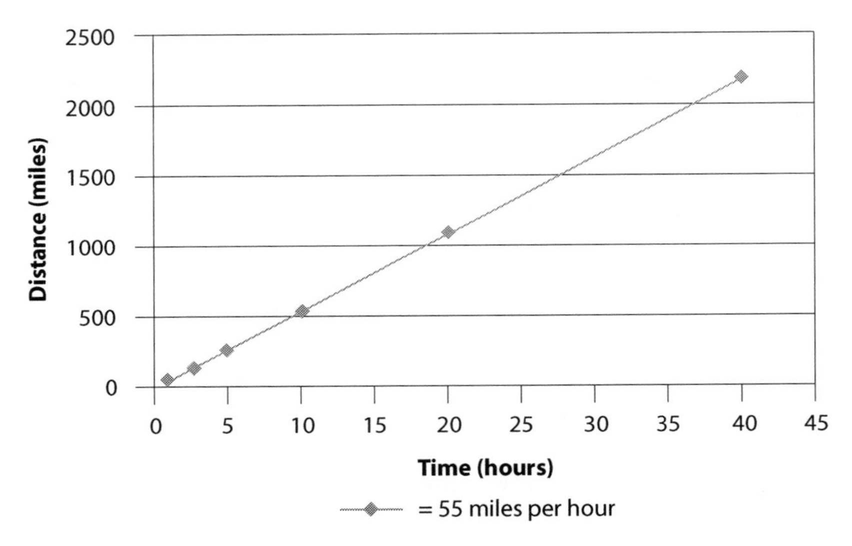

Sample answers: The graph is a straight line since the speed is constant. The line rises to the right. The slope of the line is positive.

3. Sample answer: $d = 55t$; $1900 = 55t$; $t = 35$ hours; Driving 7 hours per day is reasonable. It would take about 5 days to reach the Grand Canyon driving for 7 hours per day.

4. Sample answer: Distance traveled depends upon time spent driving. You can change the time spent driving directly but you can only change the distance indirectly (by changing the time).

5. Time is the independent variable and distance is the dependent variable. Since distance depends on time, distance must be the dependent variable.

6.

Time (hours)	Distance (miles)
1	65
3	130
5	325
10	650
20	1300
40	2600

7.

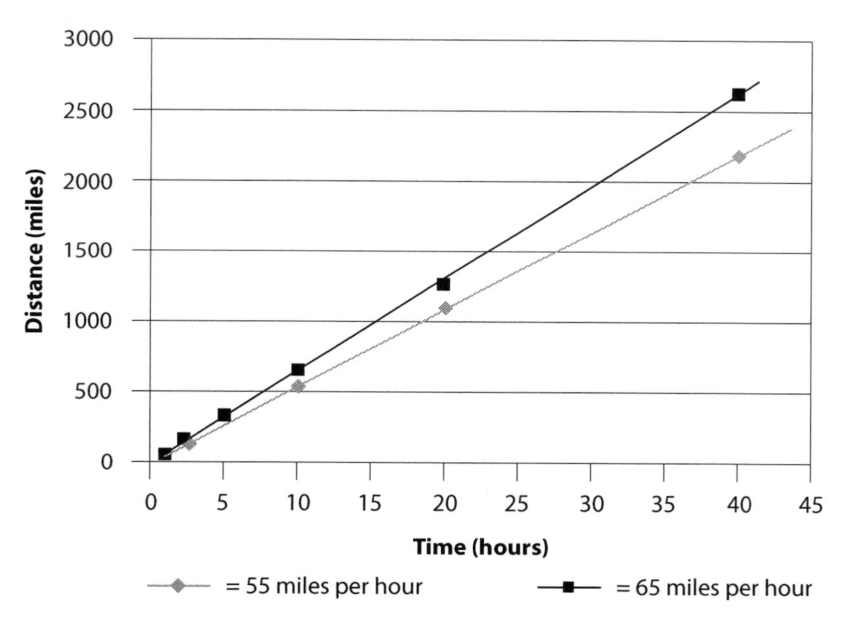

Sample answers: Both graphs are straight lines since the speed in each case is constant. Both lines rise to the right. The slope of both lines is positive. The line with the speed of 65 miles per hour is steeper than the line with the speed of 55 miles per hour. The second line (65 miles per hour) has a greater slope than the first line (55 miles per hour).

8. Sample answer: $d = 65t$; $1900 = 65t$; $t = 29$ hours; Driving 7 hours per day is reasonable. It would take about 4 days to reach the Grand Canyon driving for 7 hours per day. Driving at 65 miles per hour instead of 55 miles per hour will save about 1 day of driving.

9. The variables have not changed. Only the speed has changed.

10. No. The distance still depends on the amount of time spent driving. Time is still the independent variable and distance is still the dependent variable.

11. Sample answer: Dependent and independent variables are related to each other. As the independent variable changes, the dependent variable also changes. Other examples are time of year (independent) and average temperature (dependent), calorie consumption (independent) and weight gain/loss (dependent), and hours worked (independent) and weekly pay (dependent).

Differentiation

Some students may benefit through the use of calculators during this task; others may enjoy an opportunity to use computer software (Excel, for example) to create the table and graph. Students who complete the task early could calculate gas usage and cost for various mpg estimates.

Technology Connection

Students could create a spreadsheet to create the table and graph, or they could use a graphing calculator to compile the data and create a graph. Student can also use map software to find the exact mileage from their school to their destination.

Choices for Students

Following the introduction, offer students the opportunity to determine a different route to their chosen location and complete the task using the revised distance. They could also complete the task using different speeds.

Meaningful Context

This task makes use of a real scenario of planning a trip. The task allows students to determine how long it will take to reach a certain destination while reinforcing the ideas of independent and dependent variables.

Recommended Resources

- AAA
  www.walch.com/rr/CCTTG6TripTools
  Find tools on this site to calculate drive time and fuel costs for road trips, as well as get directions.
- Illuminations—Bouncing Tennis Balls
  www.walch.com/rr/CCTTG6TennisBalls
  This lesson plan details an activity that can be used to explore the relationship between dependent and independent variables.
- MapQuest
  www.walch.com/rr/CCTTG6MapQuest
  This site provides driving directions, distance, and estimated times. It also provides links to calculate gas usage and costs.
- TravelMath
  www.walch.com/rr/CCTTG6TravelCalculator
  Determine drive time, fuel costs, and distance with this travel calculator.
- Trip Calculator
  www.walch.com/rr/CCTTG6TripCalculator
  Select a starting point and destination on this interactive map to generate distance, driving time, and fuel costs.

NAME: ______________________________

6.EE.9.2 Task • Expressions and Equations
Planning a Trip

Part 1

You and your family are planning a road trip next summer. You get to select the destination. Your parents have asked you to help figure out the travel time and distances so they can make arrangements for hotels on the way. After selecting the destination, you will determine how far it is from your home and how long it will take to get there.

Selected destination: ______________________________

Distance: ______________________

1. You estimate that the car will travel at a speed of 55 miles per hour. Make a table showing the total miles traveled after 1 hour, 2 hours, 5 hours, 10 hours, 20 hours, and 40 hours.

2. Plot the points on a graph using graph paper. Describe the graph.

3. Develop an algebraic equation to represent the relationship between time traveled and distance traveled using the speed of 55 miles per hour. Use this equation to determine the time it will take your family to travel to your destination (round to the nearest hour). Think about a reasonable amount of time to travel in the car each day and determine how many days it will take you to get to your destination.

continued

Planning a Trip

4. Think about the relationship between the time traveled and the distance traveled. Time and distance are called variables. Can you change one of the variables without changing the other? Explain fully.

5. Think some more about the variables here. An independent variable can be changed directly while a dependent variable can be changed indirectly, or only by changing another variable first. Can you determine which variable is the independent variable and which is the dependent variable? Explain fully.

continued

6.EE.9.2 Task • Expressions and Equations
Planning a Trip

Part 2

Your parents have discovered that the speed limits on the trip will allow them to travel at a speed of 65 miles per hour.

6. Using this new speed, make a table showing the total miles traveled after 1 hour, 2 hours, 5 hours, 10 hours, 20 hours, and 40 hours.

7. Plot the points on the same graph used in Part 1. How does this graph compare to the first one? Explain fully.

8. Develop an algebraic equation to represent the relationship between time traveled and distance traveled using the speed of 65 miles per hour. Use this equation to determine the time it will take your family to travel to your destination (round to the nearest hour). Think about a reasonable amount of time to travel in the car each day and determine how many days it will take you to get to your destination. How does this compare with your calculations using a speed of 55 miles per hour?

continued

9. Think about your new equation. How does changing the speed change the equation? Have the variables changed?

10. Has the new speed changed which variable is the independent variable and which is the dependent variable? Explain fully.

11. Draft a general statement about independent and dependent variables. Give at least two other examples of independent and dependent variables.

Finding Patterns in Design

Common Core State Standard

Solve real-world and mathematical problems involving area, surface area, and volume.

6.G.1. Find the area of right triangles, other triangles, special quadrilaterals, and polygons by composing into rectangles or decomposing into triangles and other shapes; apply these techniques in the context of solving real-world and mathematical problems.

Task Overview

Background

Decomposing polygons into smaller and simpler polygons is a critical geometry skill that forms the basis for many higher-level geometric theorems. Students often struggle with seeing the shapes inside other shapes. Composing and decomposing these shapes using manipulatives assists students in seeing the connections between simpler and more complex polygon shapes.

In this activity, individuals and groups of students will use quadrilaterals and triangles of known area to construct polygon designs for consideration in a local mural. Groups will then create a model of their design to display in stations around the classroom. Then, groups will use the quadrilaterals and triangles of known area to deconstruct and calculate the area of the polygon at each station.

Prior to this activity, students should have experience with finding the area of quadrilaterals, right triangles, and equilateral triangles. Students should be familiar with the terms *length, width, base, height,* and *square inches.*

Implementation Suggestions

- Each individual student should receive one sheet of polygons. Make copies from the master pages provided at the end of this task (after the student pages).
- Groups should be composed of students with multiple polygon types.
- Students will need scissors to cut out their shapes.
- Students may use tangrams or pattern blocks, if available.
- Students will work individually to complete Part 1, in groups to complete Part 2, and either individually or in groups to complete Part 3.
- Part 2 requires that students be given multiple sheets of colored paper to present the perimeter of their design.
- Part 3 requires small stations to be set up around the room.

Introduction

Introduce this activity by displaying geometric tile and quilting patterns in the room (you may want to print images from the Web sites listed in the Recommended Resources). Ask students where they have seen similar patterns or designs. Question students on how they believe each design was composed. Copy some designs to the board or overhead and use a solid color to shade the design. Ask students if they can still determine the smaller shapes that make up the pattern.

Explain that today students will work in groups to create a polygon-based design. Before they move into groups it is necessary for them to do some work as individuals. When in groups, they will use the shapes they have been given to create a single, solid, mono-color tile pattern to be used as a repeating border around a local mural commemorating the Arab Spring freedom protests in the Middle East.

Monitoring/Facilitating the Task

- Pause when students have completed Part 1 in order to organize groups. Assign each group a location for their station. Allow groups to work and set up their stations independently.
- Pause when groups have completed Part 2. Make sure each station includes several copies of the perimeter drawing of the group's design.
- Students may work in groups or individually during Part 3.
- Encourage groups to think creatively. Remind them that they want their design to be unique and artistic.
- Encourage students to be accurate and precise when tracing the perimeter of their design for presentation at each station.
- Ask questions and prompt student thinking so that they:
  - Share and make any necessary corrections to their solutions to Part 1 when in groups.
  - Recognize when they have combined the tiles in a new polygon, such as two triangles and a square of the correct dimensions creating a trapezoid. Encourage students to recognize these polygons and use the correct math vocabulary to describe them.
  - Notice that they can calculate the area of their new polygon by summing the areas of the individual polygons used in the new shape.
  - Compose their design with consideration of the constraints they have been given.
  - Work together at each station to determine which smaller polygons were used to develop the larger design.
  - Remember to use the smaller polygon models they used to compose their own design when attempting to calculate the area of the larger designs.

Debriefing the Task

- Have each group present their design as shown in each station. They should also present the sketch they made of their design in Part 2 of the task and the area of their final design.
- Have students compare the designs each group composed. Are they similar or distinct? Encourage student discussion about the number and types of shapes that can be created from smaller polygons.
- Ask students to describe how knowing the area of the smaller shapes helped with finding the area of the larger shapes.
- Compare the student sketches done at each station. Are there situations where the same design was broken down into different combinations of polygons, yet the calculated area was the same? Encourage students to discuss why this is possible. Stress that there are multiple paths to the correct answer.

Answer Key

1. Area of quadrilaterals (smallest to largest): 0.25 in.^{2}, 1 in.^{2}, 2 in.^{2}
   Area of triangles (smallest to largest): 0.25 in.^{2}, 0.5 in.^{2}, 1 in.^{2}

2–8. Polygon designs and area computations will vary by group. Check for completion and accuracy.

Table results for Part 3 will vary based on group designs. Check sketches and area computations for accuracy.

Differentiation

Some students may benefit from the use of geometric tiles, tangrams, or pattern blocks in this task. Some may benefit from the removal of the design constraints.

Advanced students may be encouraged to compose their designs with negative space—areas where the polygon is not present. Advanced students may also be given stricter constraints for the design, such as working with smaller dimensions or using an exact total area.

Technology Connection

Designs can be created using graphic design software. Students would need to create their individual tiles according to given dimensions and then combine them into a larger design digitally. Solid colored printouts of the design could be used at each station.

Choices for Students

Students may choose to find complex polygon designs in the world. They can present the polygon design along with its smaller polygons and how they determined the area of the shape.

Students may choose to design a quilt comprised of repeating polygons as an alternative to the mural border.

Meaningful Context

Have students design and construct a mural depicting Arab Spring or another recent historical event of their choosing. Use their designs as a border of the mural.

Recommended Resources

- Arab Spring: an interactive timeline of Middle East protests
  www.walch.com/rr/CCTTG6ProtestTimeline
  This project of The Guardian news web site in London explains the events leading up to and during the Arab Spring protests. The timeline lists events up to Dec. 17, 2011.
- Democracy Now!
  www.walch.com/rr/CCTTG6ArabSpring
  This site follows developments in nations where Arab Spring protests have occurred.
- Quilt Blocks Galore!
  www.walch.com/rr/CCTTG6QuiltBlocks
  Dedicated to the design of quilt blocks, this Web site shows examples of many designs composed of smaller polygons.
- Tile Design Patterns
  www.walch.com/rr/CCTTG6TilePatterns
  This Web site shows several examples of tile combinations composed of quadrilaterals.

NAME:

6.G.1 Task • Geometry
Finding Patterns in Design

Introduction

You work for a design company that specializes in tile murals. Your job today is to use quadrilateral and triangular tiles to construct a new polygon tile design. The design you create will be repeated around the perimeter of a mural commemorating the Arab Spring freedom protests in the Middle East.

There are a few limitations on the design you create.

- The design must be created from the 6 tiles shown in Part 1 below.
- The design must be one solid color.
- The height of the design must be less than or equal to 7 inches.
- The width of the design must be less than 10 inches.
- The area of the design must be less than 70 square inches.

Part 1

Get familiar with the polygon tiles you have available.

1. Your teacher will hand out one sheet of tiles to each student. Find the area of each possible tile shape.

1/2 in. $A =$ 1/2 in.

1 in. $A =$ 1 in.

1 in. $A =$ 2 in.

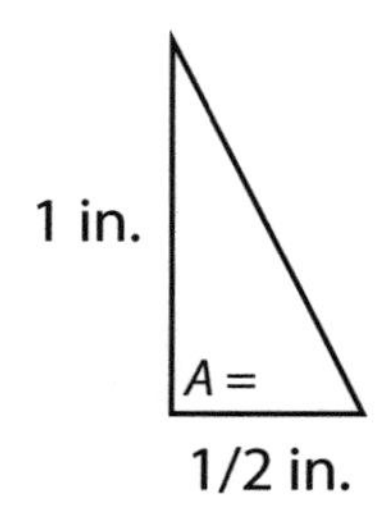

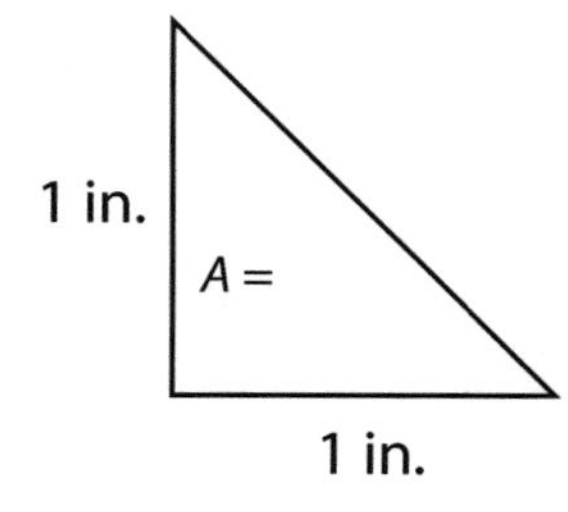

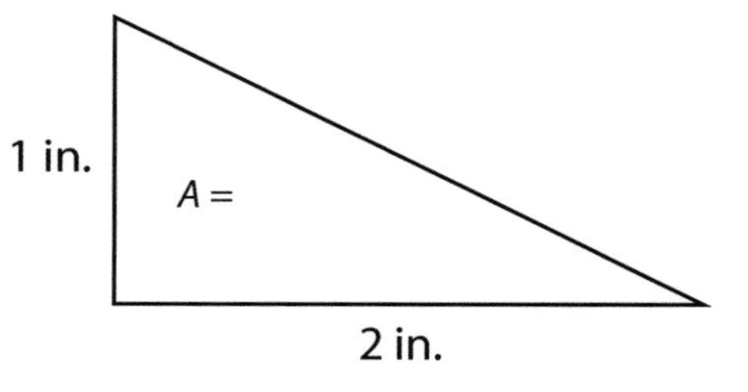

continued

2. Cut out the polygons you have been given. Be as accurate and precise as possible. Write the area on each polygon.

3. Use two or more of the polygons you have been given to construct a new polygon. Trace the polygon below. Draw in the shapes the polygon is composed of.

4. Find the area of your new polygon.

Part 2

Now that you have become familiar with your tiles, join your group to come up with the mural border design. Refer to the list of limitations under Part 1.

5. Your group should have several polygons to work with. As a group, use as many polygons as necessary to construct your design.

6. Trace the design below. Draw in the shapes that make up the design.

7. Find the area of the design.

8. Each member of the group should trace only the perimeter of the design onto construction paper. Display the perimeter design at your group's station.

continued

NAME:

6.G.1 Task • Geometry
Finding Patterns in Design

Part 3

Visit the stations in the classroom. Fill in the table below.

Station 1	**Station 2**
a. Sketch the outline of the design. b. Add lines to your sketch to show the smaller tiles that make up the design. c. Find the area of the design.	a. Sketch the outline of the design. b. Add lines to your sketch to show the smaller tiles that make up the design. c. Find the area of the design.
Station 3	**Station 4**
a. Sketch the outline of the design. b. Add lines to your sketch to show the smaller tiles that make up the design. c. Find the area of the design.	a. Sketch the outline of the design. b. Add lines to your sketch to show the smaller tiles that make up the design. c. Find the area of the design.

continued

NAME:

6.G.1 Task • Geometry
Finding Patterns in Design

Station 5	**Station 6**
a. Sketch the outline of the design. b. Add lines to your sketch to show the smaller tiles that make up the design. c. Find the area of the design.	a. Sketch the outline of the design. b. Add lines to your sketch to show the smaller tiles that make up the design. c. Find the area of the design.

NAME:

6.G.1 Task • Geometry
Finding Patterns in Design

$A =$	$A =$	$A =$	$A =$
$A =$	$A =$	$A =$	$A =$
$A =$	$A =$	$A =$	$A =$
$A =$	$A =$	$A =$	$A =$
$A =$	$A =$	$A =$	$A =$

NAME:

6.G.1 Task • Geometry

Finding Patterns in Design

$A =$	$A =$	$A =$	$A =$
$A =$	$A =$	$A =$	$A =$
$A =$	$A =$	$A =$	$A =$
$A =$	$A =$	$A =$	$A =$
$A =$	$A =$	$A =$	$A =$

NAME:

6.G.1 Task • Geometry
Finding Patterns in Design

$A =$	$A =$	$A =$
$A =$	$A =$	$A =$
$A =$	$A =$	$A =$
$A =$	$A =$	$A =$
$A =$	$A =$	$A =$

NAME:

6.G.1 Task • Geometry
Finding Patterns in Design

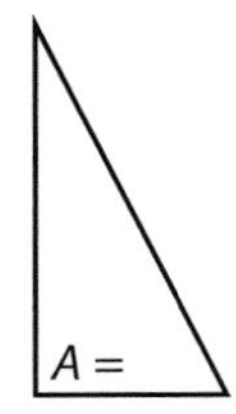

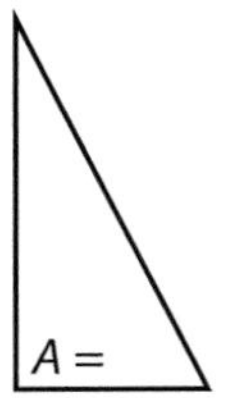

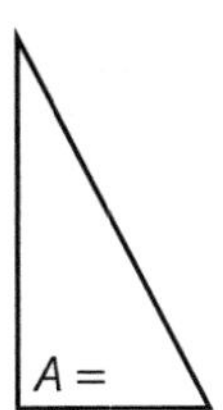

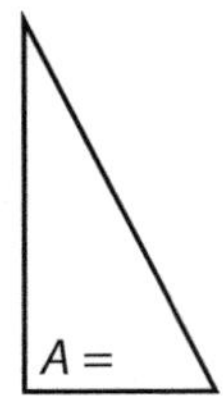

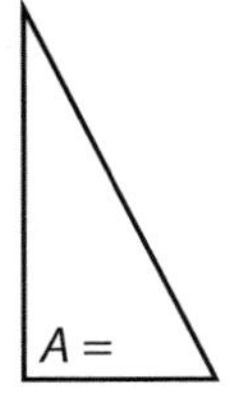

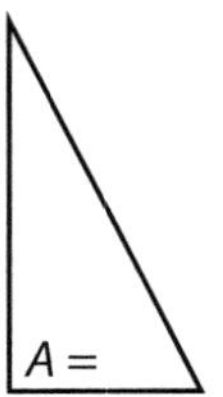

$A =$

NAME:

6.G.1 Task • Geometry
Finding Patterns in Design

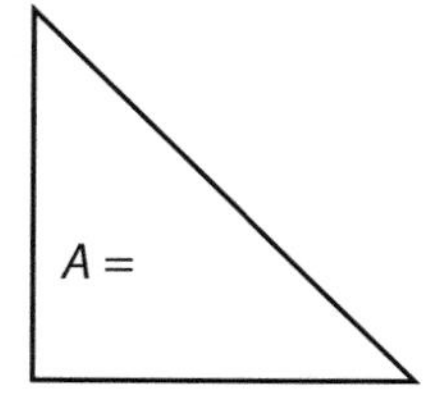

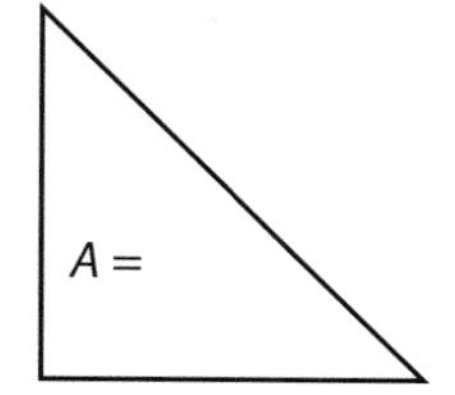

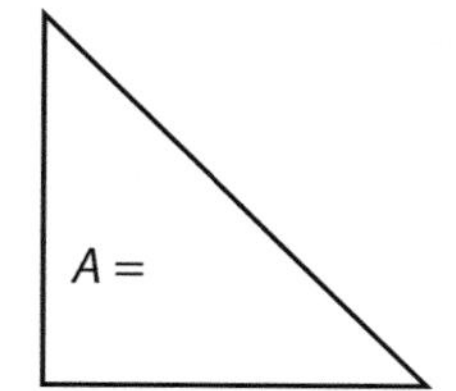

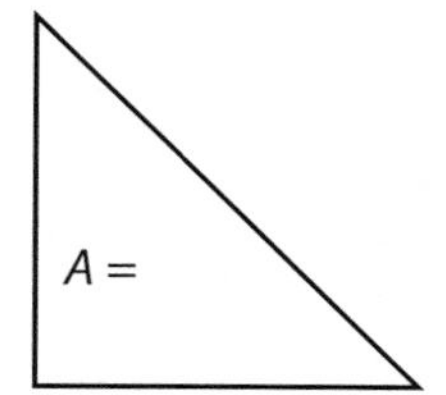

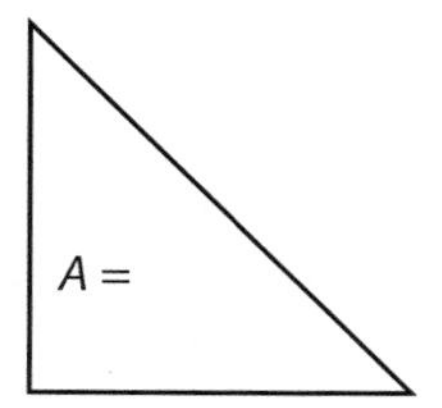

A =

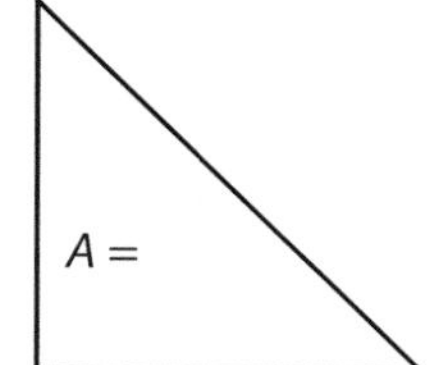

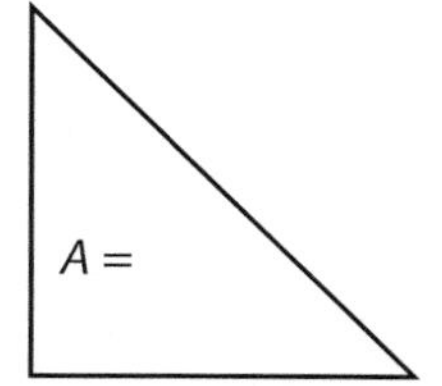

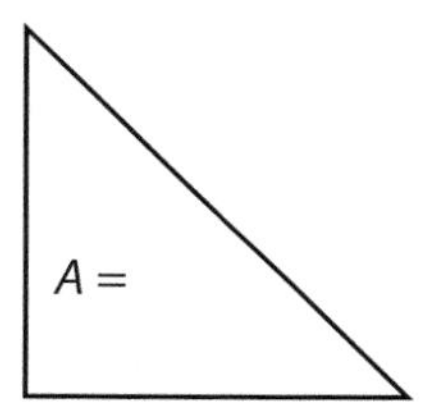

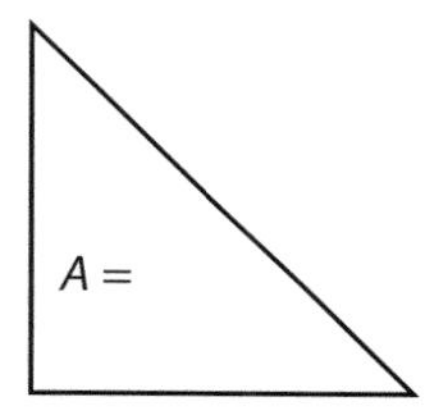

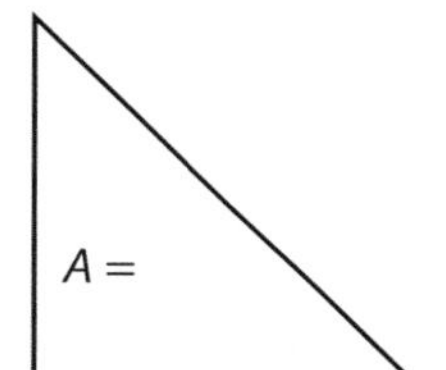

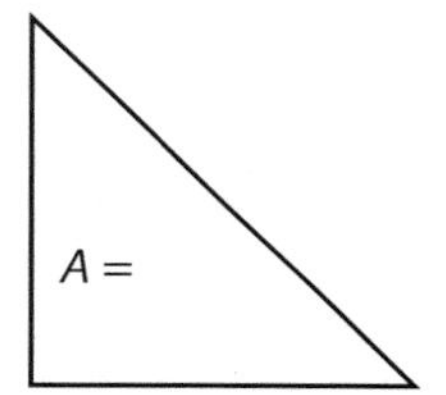

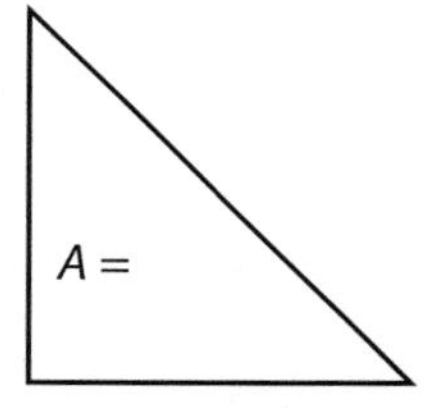

A =

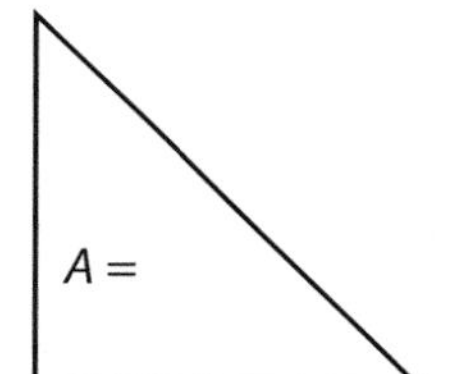

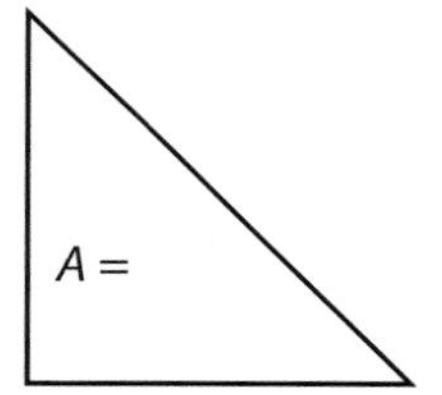

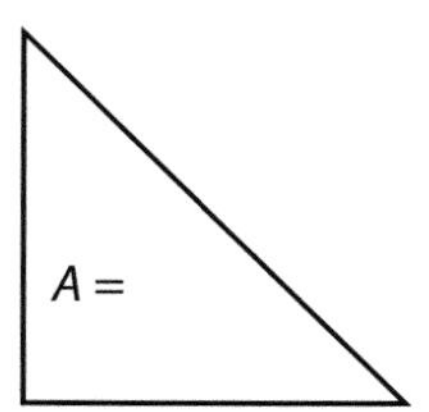

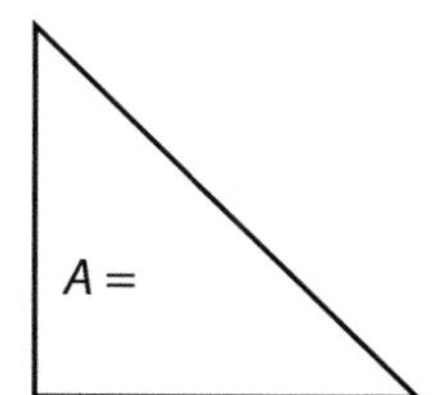

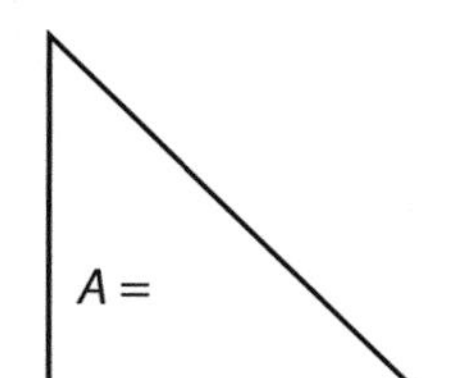

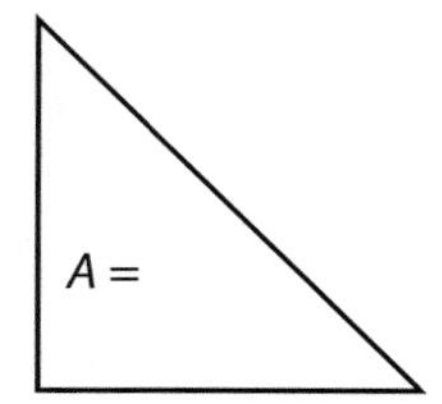

NAME:

6.G.1 Task • Geometry
Finding Patterns in Design

A =

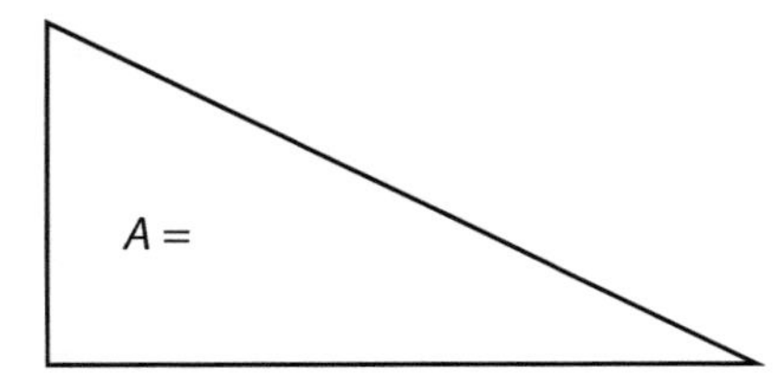

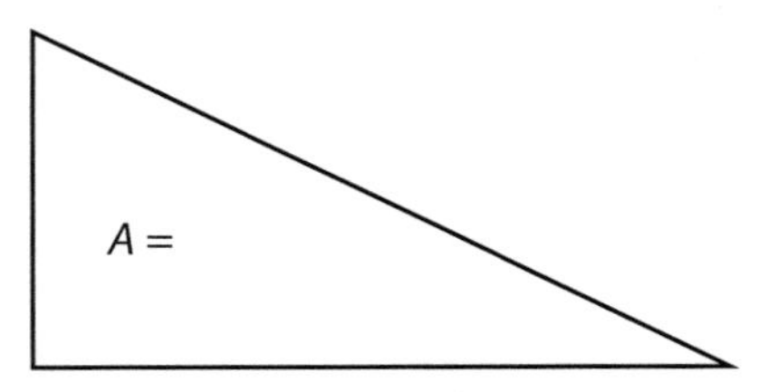

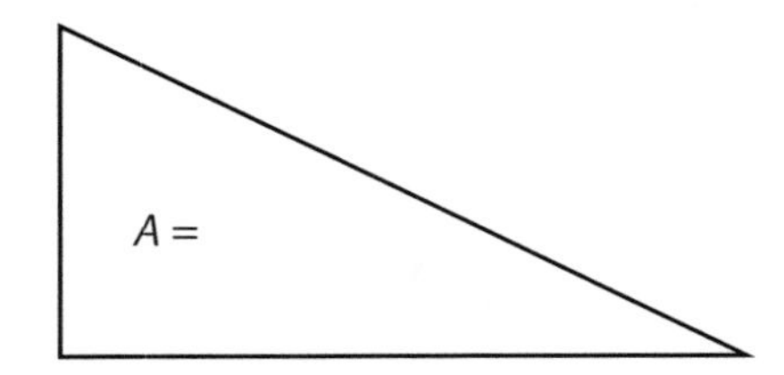

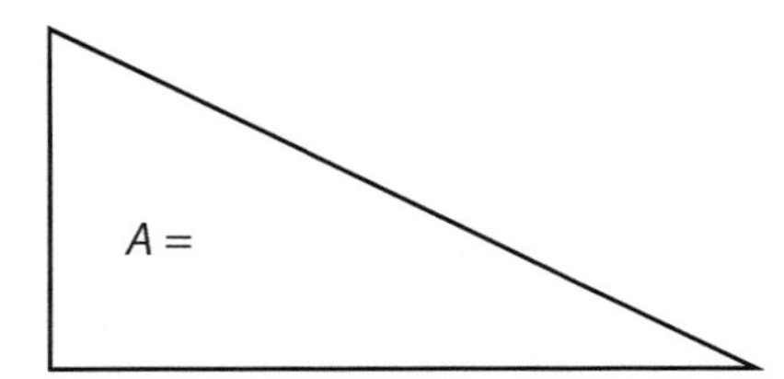

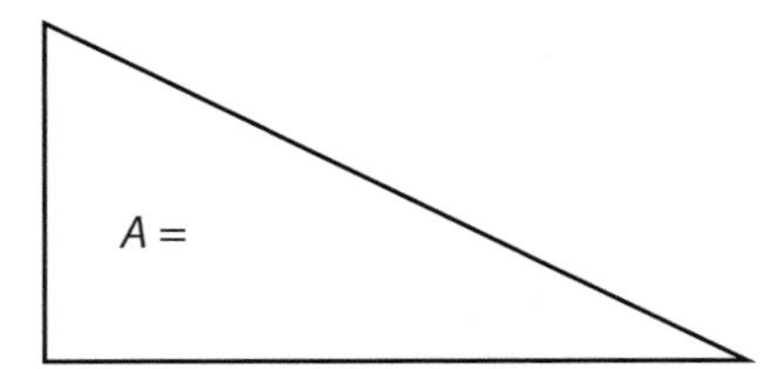

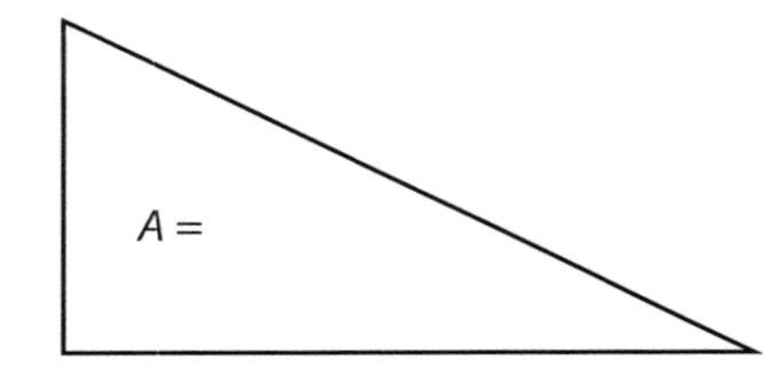

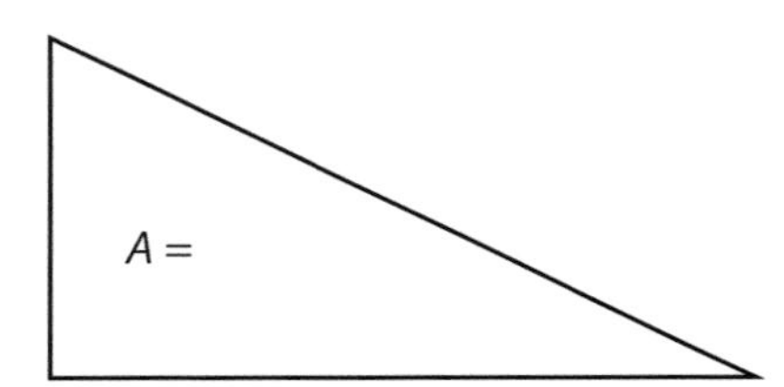

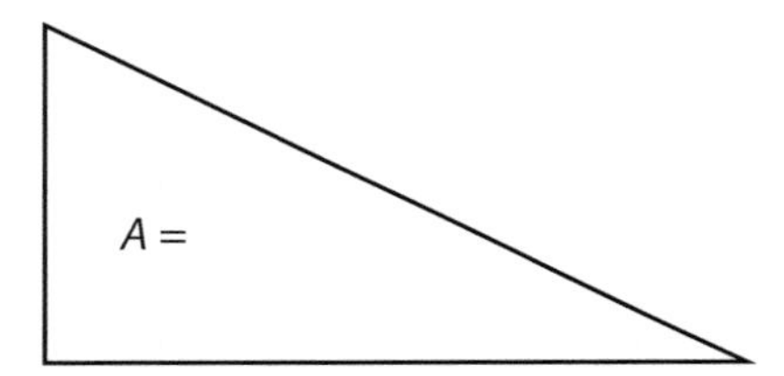

A =

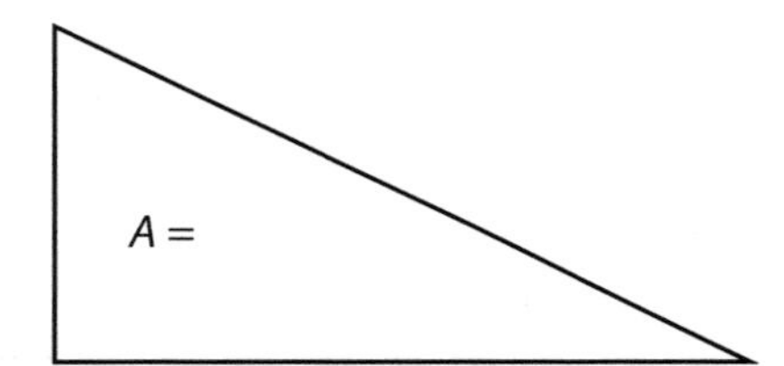

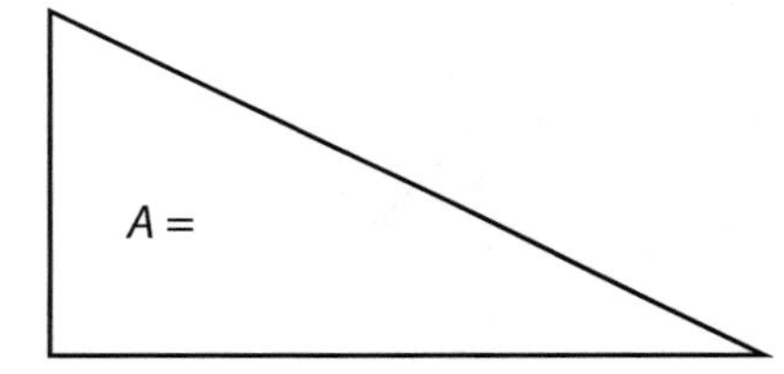

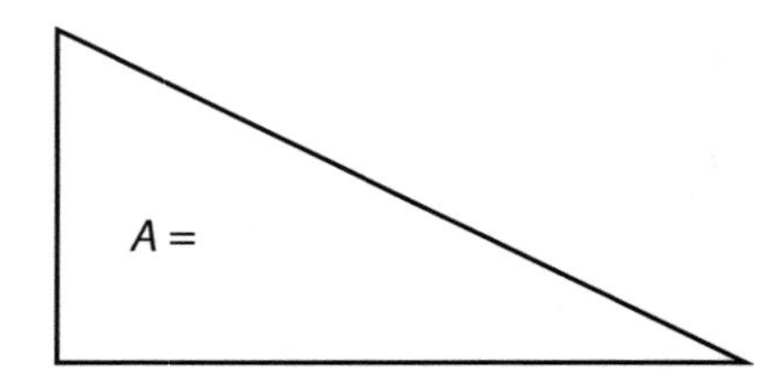

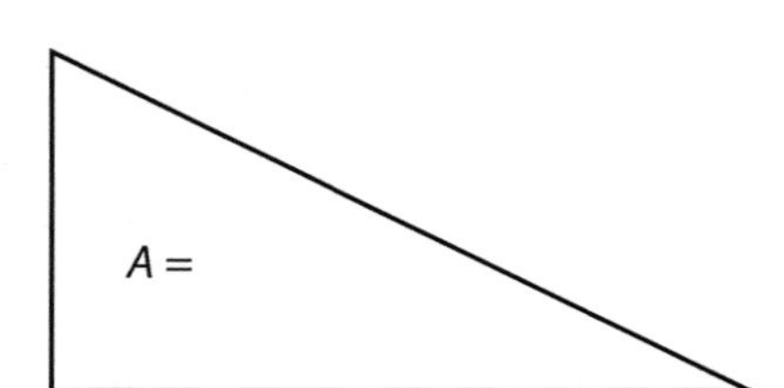

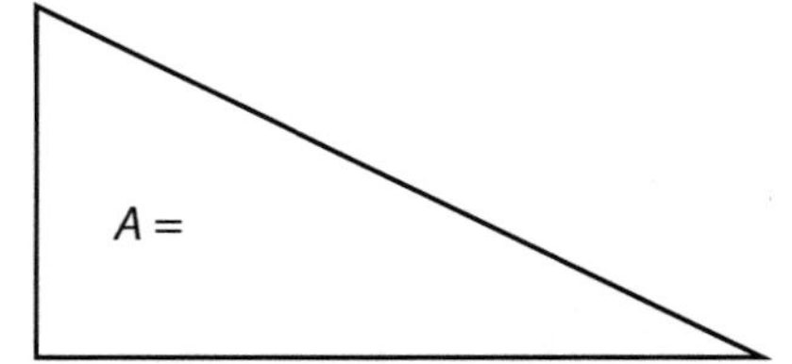

Painting the Room

Instruction

Common Core State Standard

Solve real-world and mathematical problems involving area, surface area, and volume.

6.G.4. Represent three-dimensional figures using nets made up of rectangles and triangles, and use the nets to find the surface area of these figures. Apply these techniques in the context of solving real-world and mathematical problems.

Task Overview

Background

Being able to sketch 3-dimensional drawings and find the surface area of 3-dimensional figures is a common real-life application of geometry. Many professionals such as architects, carpenters, engineers, and interior designers use these skills in their jobs.

In order to complete this task, students should be able to recognize and name 3-dimensional figures and recognize and name 2-dimensional figures. Students should also know how to find the area of common 2-dimensional figures such as squares, rectangles, and triangles. Students need to be able to multiply decimal numbers.

This task is designed to help students visualize 3-dimensional figures by drawing and analyzing their corresponding nets. Surface area can then be found by adding the areas of the individual parts of the net. A hands-on manipulative is used to help the students "see" how the net is formed. Then students are asked to visualize the net of a 3-dimensional object.

The task also provides practice with:

- measuring
- multiplying decimal numbers
- finding areas of 2-dimensional figures
- recognizing and naming 3-dimensional figures

Implementation Suggestions

Students may work individually, in pairs, or in small groups to complete one or both parts of the task.

Empty boxes will be required for this task. Individual cereal boxes or frozen waffle boxes will work well. One empty box can be shared among a group of students to conserve materials, but students should perform their own measurements and create their own drawings.

Some students may benefit from the use of isometric dot paper for the drawings.

Introduction

Introduce the task by asking students if they have ever helped paint a room in their house. Did they help determine how much paint was needed? How did they figure that out?

Begin the task by having students recall the area formulas of common 2-dimensional figures such as rectangles, squares, and triangles.

Make sure students know how to accurately measure using a ruler. If necessary, review how to measure to the nearest 0.5 inch and/or how to multiply decimal numbers.

Prepare students for group work. Make sure they have identified who will be cutting up the empty box. Each student should measure the flattened-out box. Students should create their own drawings.

Group students appropriately and provide each group with materials to complete the task.

Monitoring/Facilitating the Task

Ask questions and prompt student thinking so that they:

- Appropriately cut the empty box so the net is formed.
- Accurately measure the flattened-out box to the nearest 0.5 inch.
- Correctly sketch the 2-dimensional drawings. Some students may benefit from the use of isometric dot paper.
- Perform multiplication of decimal numbers, if applicable.
- Sketch the 3-dimensional room and properly label the sides.
- "See" the net of the room. Some students may need to refer to the manipulative to help draw the net of the room.
- Think of other real-world applications in which nets could be used.

Debriefing the Task

- Upon completion of the task, students should share their results. Make sure all students have measured the manipulative and created their own drawings. If different-sized boxes are used, answers will vary; however, students who used the same box should have similar answers.
- Ask students to share how they "saw" the net after drawing the 3-dimensional sketch. Did anyone need to refer to the manipulative in order to complete this part of the task?
- Ask students to share any difficulties they had with the task and how they overcame those difficulties.

- Encourage students who used Google SketchUp to share their drawings and explain the steps they used to create the drawings.
- Prompt students to identify the differences between this situation (having a room with no windows and painting all four walls, the ceiling, and the floor one color) and what might be more likely.
- Ask students to explore other real-world applications where nets could be used to determine surface area.

Answer Key

1. Answers will vary; however, if empty cereal boxes are used, the figure is a rectangular prism. The sides are rectangles or perhaps squares.
2. Check to make sure the box has been cut appropriately.
3. Net drawings will vary. Sample drawing (dashed lines represent the folds):

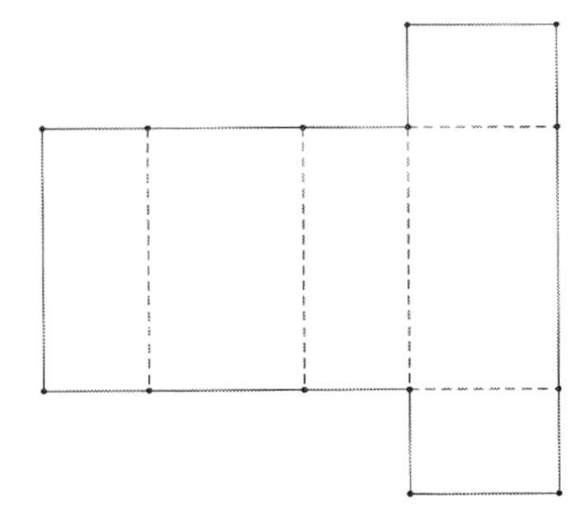

4. Answers will vary.
5. Answers will vary.
6. Answers will vary. The total surface area is the sum of the areas of the individual rectangles.
7.

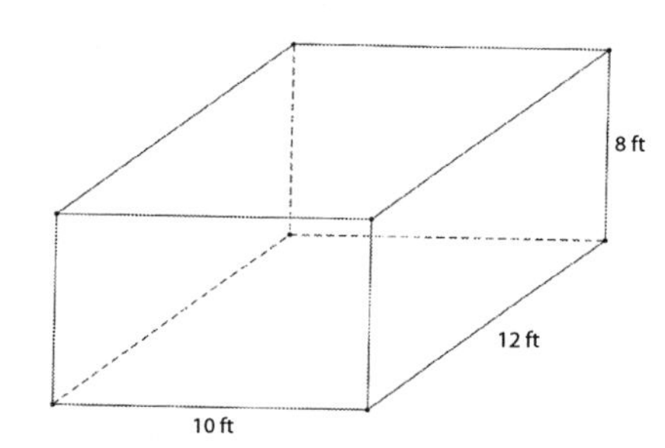

8. Net drawings may vary depending on how the students "cut up" the figure. Sample drawing (dashed lines represent the folds):

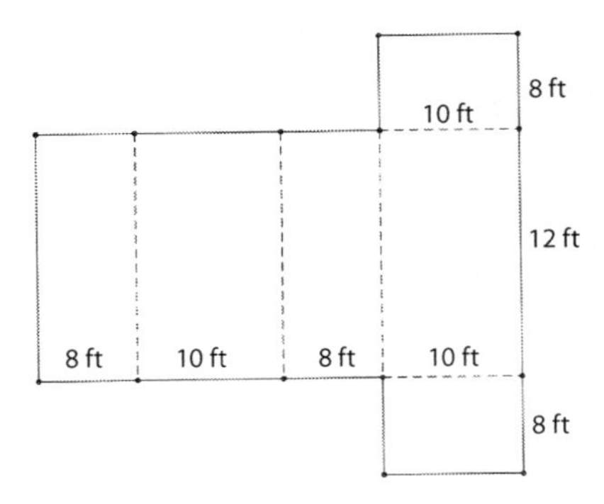

9. Find the areas of the individual rectangles. There are two $8' \times 10'$ rectangles (160 ft^2), two $10' \times 12'$ rectangles (240 ft^2), and two $8' \times 12'$ rectangles (192 ft^2). The sum of all six rectangles is 592 ft^2. Therefore, the total surface area is 592 ft^2.

10. Divide 592 ft^2 by 350 square feet per gallon of paint. The result is 1.69 gallons. Since it is not possible to buy a part of a gallon of paint, the answer should be an integer. Two gallons of paint will be needed.

11. Sample answers: calculating the amount of fabric needed to cover a box, calculating the amount of siding needed to side a house, or calculating the size of a label needed to label a box

Differentiation

Some students may benefit from using calculators during this task. Allow students to use drawing software (e.g., Google SketchUp) to create the 3-dimensional drawing. Allow students to use isometric dot paper to create the 3-dimensional drawing and the net drawings.

Students who complete the task early could research paint prices and calculate how much it would cost to buy the paint for the room. Alternatively, they could sketch a 3-dimensional figure that contains triangles and sketch the corresponding net drawing.

Technology Connection

Students could use drawing software (such as Google SketchUp) to create the 3-dimensional drawing.

Choices for Students

Following the introduction, offer students the opportunity to use different dimensions for the room and complete the task using these dimensions.

Meaningful Context

This task makes use of a real scenario of calculating how much paint is needed to paint a room. The room is not necessarily "typical" but the scenario is realistic. The task allows students to use prior knowledge of area of rectangles to determine surface area of a 3-dimensional figure.

Recommended Resources

- Cube Nets
  www.walch.com/rr/CCTTG6CubeNets
  In this activity, students determine which net would form a cube when folded up.
- Google SketchUp
  www.walch.com/rr/CCTTG6SketchUp
  This site provides free downloadable drawing software, which can be used to make 3-dimensional drawings.
- Home Depot
  www.walch.com/rr/CCTTG6HomeDepot
  Students can search this site for paint prices.
- Isometric Dot Paper
  www.walch.com/rr/CCTTG6DotPaper
  Printable isometric dot paper is available from this site.
- Shape and Space in Geometry
  www.walch.com/rr/CCTTG6Geometry3D
  This collection of interactive lessons promotes visualization of 3-dimensional objects.

6.G.4 Task • Geometry
Painting the Room

Part 1

Sam's parents have agreed to let him move to a bedroom in the finished basement, away from his annoying little brother! However, the room isn't ideal—it has no windows, a concrete floor, and a stained ceiling. His parents are letting him paint the room (all four walls, the door, the floor, and the ceiling) whatever color he wants! Now he just needs to pick a color and figure out how much surface area there is in the room. In this task, you will help him find the surface area of his room.

Use nets made up of rectangles or squares to find the surface area of 3-dimensional solids. The room is a box—a 3-dimensional solid. First you will make a net drawing of a 3-dimensional solid. Then you will calculate how much paint Sam will need to paint the inside of his room.

1. Look at your empty box. What kind of 3-dimensional solid is it? Describe the 2-dimensional figures that make up the sides, top, and bottom of the figure.

2. Cut the dashed edges of the box as shown. When you are done cutting, the box should still be in one piece and you should be able to lay it out flat.

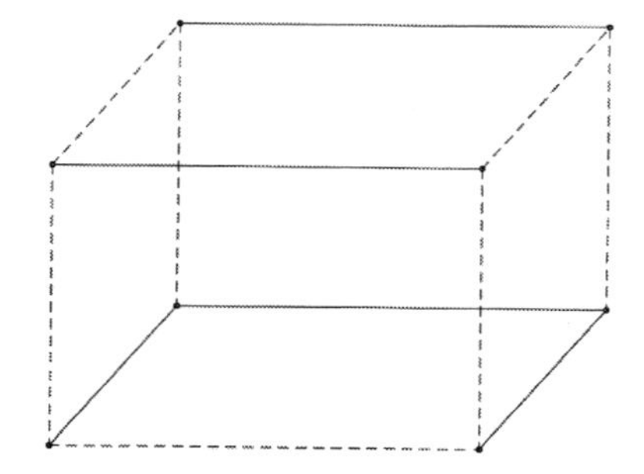

continued

3. Sketch a drawing of your flattened-out box. This drawing is called a net.

4. Using a ruler, measure the sides of the box. Round to the nearest 0.5 inch.

5. Identify the figures that make up the net. How can you determine the area of each figure?

6. How can you determine the total surface area of the box? Explain your steps.

continued

NAME:

6.G.4 Task • Geometry
Painting the Room

Part 2

You will now use the concept of nets to determine the surface area of Sam's room. Then you can calculate the amount of paint he will need. The room has no windows and one door. Remember, he gets to paint the walls, door, ceiling, and floor all the same color. The room is $10' \times 12'$ and the ceiling is $8'$ high.

7. Sketch a 3-dimensional drawing of the room. Appropriately label the sides.

8. Now sketch a net drawing of the room. Be sure to label the sides appropriately.

9. How can you use the net drawing to find the total surface area of the room? Describe the process and show your steps.

continued

6.G.4 Task • Geometry
Painting the Room

10. If a gallon of paint covers 350 square feet, how many gallons of paint will Sam need? Describe the process and show your steps. Think about what type of number (fraction, integer, or decimal) your answer should be and why.

11. Think about the use of nets. Can you think of another example in which nets could be used to find the surface area of a 3-dimensional solid? Describe fully.

How Old Were You When . . .?

Common Core State Standard

Develop understanding of statistical variability.

6.SP.2. Understand that a set of data collected to answer a statistical question has a distribution which can be described by its center, spread, and overall shape.

Task Overview

Background

In this task, students develop a statistical inquiry based on the question, "How old were you when ...?" Students will collect data, summarize their results in a histogram, and present their findings to the class. Discussion will then focus on comparison of the data sets, especially in relation to spread, center, and shape.

Prior to this task, students should be familiar with data collection and analysis. They should have used tables and graphs to record and present data. They should be familiar with common statistical vocabulary—sample size, spread, range, center, shape, and average.

This task also provides practice with:

- phrasing statistical questions
- collecting and organizing data
- presenting data in a graphic format
- calculating summary statistics for a data set

Implementation Suggestions

Please note: Defining questions, collecting, organizing, and displaying data are precursors to the focus of the standard—the analysis of data sets. As much as possible, streamline the data collection and graphing. Guide students through quick, efficient question formulation and data collection. Consider assigning question formulation as homework OR doing data collecting and graphing on one day and the debrief during a second class period.

Students may work individually or in groups for this activity. Consider organizing students into a "reverse receiving line" for data collection. Have them line up shoulder to shoulder. The student at the front of the line should survey each student as he or she walks along the line, and then join the line at the end. The next student in line then surveys everyone including the first student and joins the end of the line. The procedure is repeated until the entire line is cycled through.

Emphasize organization and display of data to support data analysis.

Presentation materials should be made available during this task. Preparation of a display area for posters and space available for a poster walk may ease the sharing of data.

Introduction

Ask students to think about how many people are in the United States. Ask if they could imagine having to ask so many people the same question, record their answers, present their responses, and understand what the responses mean. Have students suggest ways to survey an entire nation of people.

Ask students how we make sense of the data collected. If necessary, review the concepts of spread, center, and shape as they relate to a data distribution.

Explain that today the class will become a mini census bureau, surveying each member of the class, and then recording, organizing, displaying, and analyzing the data. Ask students to think of "firsts" in their lives: their first trip to the state capital, their first day of school, their first sleepover. This will help them brainstorm ideas for their survey question.

Monitoring/Facilitating the Task

Ask questions and prompt student thinking so that they:

- Format their question in the standard form. Students may focus on yes/no questions. Help them transition their question to one that will elicit a quantitative answer, e.g., "How old were you when you first ...?"
- Organize their data collection effectively. Provide a class list and/or encourage students to develop a table or tally system to record their results and to keep a record of who they have surveyed.
- Consider how they will record responses of "I have never done that." Allow students to consider if they should or should not record that data and act accordingly.
- Create a clear and effective graph of their data. Students should use a bar graph, histogram, or scatter plot to display their data. Make sure each axis is labeled and that the unit markers used are logical.
- Properly calculate the mean, median, mode, and range of their data.
- Consider the spread of their data by asking if their graph is shaped like a square, an arc, two separate buildings, or a mountain range. Encourage them to be descriptive.
- Use comparison terms such as *less/more, larger/smaller,* and *wider/thinner* when they are describing their data.
- Prepare a brief presentation that covers the required elements.

Debriefing the Task

- Have students present their work or display student work in the front of the classroom.
- Allow students to view one another's work. Ask students to focus on the **distribution**—the center, range, and shape of the data their classmates collected. If desired, and if time allows, discuss data collection strategies and/or the purpose of anonymous surveys.
- Review and discuss the data collected by the class. Draw students' attention to the **distribution** or "**shape**" of the data.
- Ask students to remark on interesting questions, patterns, and any other observations they made.
- Discuss the **center** (mean, median, mode) of each data set.
- Consider which questions led to a smaller (younger) center and vice versa.
- Ask students about how well the **center** represents the distribution of the data. Relate this to the **range** of the data. Note that a data set with a small range had a center value that's more representative than a data set with a large range.
- Have students describe the shapes of the graphs. Discuss the normal (bell curve), uniform (rectangular), and binomial (two buildings) distributions. Ask students to think about why certain questions would yield answers of different distributions. For example, "How old were you when you entered school?" could be binomially distributed with students primarily starting school at age five or six. "How old were you when you first flew in a plane?" may be more uniformly distributed, as a first flight may occur at any time in a person's life.
- If possible, point out a survey with a low sample size. Ask the students why the sample size might be lower when all the students were surveyed. Ask students about the reliability of data collected with a small sample size (lots of "n/a" responses or a small population).

Answer Key

Answers will vary but mean, median, and mode should be correctly calculated, and center, spread, and overall shape should be accurately described based on the data.

Important assessment will take place during the debrief. Gauge students' perceptions of the data sets, including the sets' similarities and differences, and the implications of each set's "shape" or distribution.

Differentiation

Some students may benefit from having a partner during this task. Some students may benefit through the use of a graphic organizer for data collection.

Technology Connection

Students may enter their data into a spreadsheet and do their calculations and graphing electronically.

As an extension, use a Web site such as "Post Your Info" (see Recommended Resources for URL) to research other statistical questions and results. Student-collected data can be compared to this data or combined with it, providing a larger sample size for analysis. Students can observe similarities and differences between their classroom and the data collected from worldwide Internet users. These observations can lead to deeper analysis of the statistics or a more extended interdisciplinary project.

Choices for Students

Students can develop their own question outside of the formatted one for this task. It is important that the question yield quantitative data for the calculation of summary statistics and analysis of center, spread, and shape.

Meaningful Context

Students can design a larger survey based on the question, "How old were you when ...?" Have students decide on a question they would like to ask their families or schoolmates. The question may relate to their community ("How old were you when you moved to your current neighborhood?") or historical events ("How old were you on September 11, 2001?"). Students can make connections with their families and friends by collecting data and exploring their statistical question with a larger sample size. Students could present the data to the school and/or their families following their analysis of center, spread, and shape.

Recommended Resources

- Measure of Central Tendency
  www.walch.com/rr/CCTTG6CentralTendency
  This site provides a review of mean, median, and mode.
- Post Your Info
  www.walch.com/rr/CCTTG6FamilyStatistics
  This site allows Internet users to input their responses to questions relating to family, home, self, and other topics. Some questions are in the form, "How old were you when...?" The site presents data in multiple forms and provides some summary statistics and tools for analysis.
- U.S. Census Bureau
  www.walch.com/rr/CCTTG6CensusBureau
  www.walch.com/rr/CCTTG6AboutCensusBureau
  The U.S. Census Bureau collects, analyzes, organizes, and provides data about the nation's people and economy. The first link above provides information about the U.S. Census and links to the data collected in current and past years. The second link provides specific information on the purpose of the work the bureau completes each year.

6.SP.2 Task • Statistics and Probability
How Old Were You When . . .?

Part 1

The U.S. Census Bureau collects data from and about all the people in the United States. That data is used for a number of purposes, from assigning Congressional seats to distributing neighborhood funding. Today you will design and participate in a statistical study of your classmates.

1. First, write a question to ask your classmates:

   "How old were you when ____?"

2. Next, collect the data. Use the box below, or a class list provided by your teacher, to record your data. Make sure to survey everyone in the classroom.

3. Briefly describe how you collected your data and ensured that you surveyed everyone and recorded the data accurately.

continued

4. Organize your data. Create a table in the space below to organize the data you collected.

5. Graph your data. Use the table you created to complete the graph below. Make sure to put a title on your graph and label each axis appropriately.

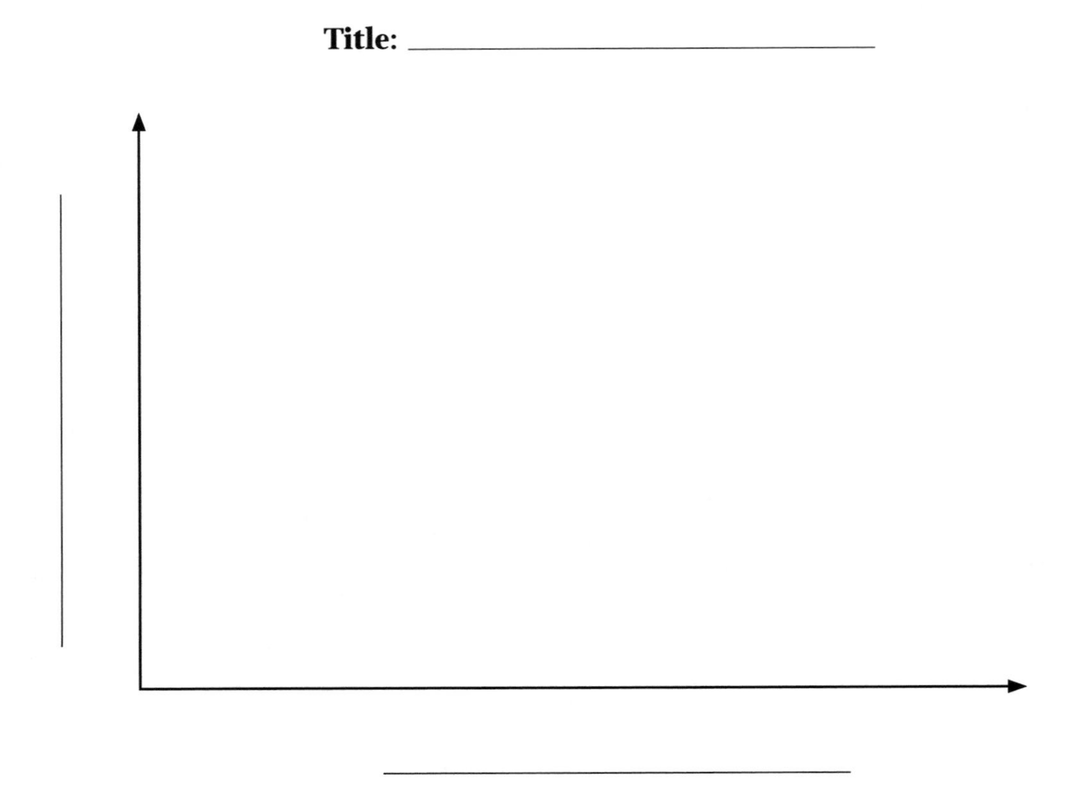

continued

6. Describe what the graph shows.

7. Summarize your data. Calculate summary statistics for your data set.
   a. Calculate the mean age.
   b. Calculate the median age.
   c. Calculate the mode age.
   d. Calculate the range.

8. Analyze your data. Think about what your data is showing you about your question.
   a. Describe the spread of your data.
   b. Describe the center of your data.
   c. Describe the shape of your data.

continued

6.SP.2 Task • Statistics and Probability
How Old Were You When . . .?

Part 2

Now you must prepare your data and analysis for presentation to other people. This needs to be neat, organized, and easy for your audience to understand.

Make sure your presentation includes:

- the question you posed
- the sample size of the people you polled
- a visual representation of your results
- the summary statistics you calculated
- an analysis of the spread, center, and shape of your data

Be prepared to share your presentation with your classmates.

Water Flowing into the Chesapeake Bay

Common Core State Standard

Summarize and describe distributions.

6.SP.4. Display numerical data in plots on a number line, including dot plots, histograms, and box plots.

Task Overview

Background

How does the average amount of water flowing into the Chesapeake Bay vary from year to year and decade to decade? In this task, students will use publicly available data to summarize and compare mean annual streamflow into the Chesapeake Bay over the past seven decades.

Box-and-whisker plots are most often used in the classroom to display information related to student grades. This task relates box-and-whisker plots to their most common real-life application. Box-and-whisker plots are regularly used in scientific study to compare two or more data sets.

The task also provides practice with:

- calculating the median, first quartile, third quartile, maximum, and minimum of a data set
- making conjectures about factors contributing to streamflow variation over time
- reasoning instinctively about data
- constructing arguments based on data
- using, analyzing, and interpreting data tables and data summaries

Implementation Suggestions

Students should be divided into seven groups (one group for each decade). An eighth group could be assigned the entire data set to work with. Alternatively, the data set could be divided in time periods of 15 or 20 years for classrooms with fewer groups.

Go to www.walch.com/CCTTG6WaterData to find Chesapeake Bay streamflow data. The entire data table and summary graphic should be available to each student.

Introduction

The task should be implemented following instruction on the mechanics of creating a box-and-whisker plot. Students should be familiar with, and able to calculate, the median, maximum, minimum, and quartiles of a data set.

Instruction

Begin the task by having students read and familiarize themselves with the streamflow information they have been given. Have the students complete Part 1. Discuss the answers as a class. Students should observe that the data is an annual mean calculated using the mean monthly streamflow for each year.

Prepare students for the group work. Explain that each group will be creating a box-and-whisker plot for display in the classroom. Each group will be assigned a decade from the information table they have just discussed. Explain that their box-and-whisker plots will then be used to compare streamflow over different decades.

Have student groups determine standards for their own box-and-whisker plot displays. Provide guidance to make sure that they consider the use and format of a title, the size of the box-and-whisker plot, the placement and wording of the data citation, and the scale and unit values for the horizontal axis. Prompt students to think about the fact that the horizontal axis should cover the range of the entire data set and unit markers should be consistent for the comparison done during the debriefing.

Divide students into groups and assign each group a time period (e.g., 1941–1950, 1951–1960, etc.). Depending on classroom norms, you may want to assign roles within the groups. Possible roles include the facilitator, the graphic designer, the calculator, the checker, and/or the timekeeper. Assigning roles provides assurance that each student in the group participates in the process, and it provides a way of monitoring individual participation.

Monitoring/Facilitating the Task

While groups are working, ask questions and prompt student thinking so that they:

- Accurately calculate the values necessary to complete each box-and-whisker plot. Group members should each be calculating and checking values. It may be helpful to create the box-and-whisker plot in preparation for this activity. *Note:* An answer key is not provided as year breakdowns may be varied and quartile calculation methods may differ. See the Recommended Resources for more information on methods of quartile calculation.
- Justify their answers and their methods and reflect on their answers as compared to others.
- Explore the summary data using Part 3 of the worksheet, after completing their box-and-whisker plot. They should examine the range (spread) between the maximum and minimum and quartile values for streamflow. Make sure students understand that 50 percent of the data falls between the first and third quartiles. Focus on the question: How does the average amount of water flowing into the Chesapeake Bay vary from year to year? Ask students to justify their responses using their box-and-whisker plot.
- Make connections between what their box-and-whisker plot may indicate and precipitation amounts for their decade.

Note: Some groups may need additional support to effectively answer the focus question. This support can be provided verbally during the presentations or designed into a worksheet to be completed during group time. Make sure to:

- Encourage deeper comprehension of the box-and-whisker plot in series, drawing the student's attention to the summary values and strengths of the box-and-whisker display.
- Focus students on the range between maximum and minimum streamflow values and the range between the first and third quartiles so that they see that wide range indicates variability; a smaller range indicates that most of the years were similar in their streamflow amounts.

Debriefing the Task

Upon completion of their box-and-whisker plot, each group should present their results to the class. Their presentation should include their decade, their calculations, and an explanation of their box-and-whisker plot. They should also cover their response and justification to the focus question. Allow the individual groups to present their information and for the remaining groups to question the presenters. Make sure that the calculations and box-and-whisker plot presented are accurate.

Upon completion of their presentation, each group should display their box-and-whisker plot for comparison between decades. Students should be able to view and compare each plot. Aligning their work vertically in order of decade is an effective method to do this.

As a whole class, expand on the focus question to encourage students to begin thinking about streamflow from decade to decade. How does the average amount of water flowing into the Chesapeake Bay vary from decade to decade? Use the visual display to compare median streamflow, the interquartile range, and the maximum and minimum for each decade.

Ask students to create conjectures in regards to precipitation amounts from decade to decade. Remind them to use the plot to justify their ideas and responses. Students should suspect that lower streamflow years followed years with low precipitation, and vice versa.

Possible prompts/questions include:

- Identify differences among box plots and make conjectures as to why those differences may occur.
- Did any groups have the same box-and-whisker plot but different sets of numbers?
- Why are the boxes different shapes?

After presenting the prompts/questions, facilitate a class discussion about the different components and what affects the position and size of each section.

Answer Key

1. water year, annual mean streamflow to the bay, normal/above/below average flow
2. Calculated. The information above the table indicates that the streamflow information is calculated using monthly mean data.
3. Chesapeake Bay, Maryland
4. 1937 through 2010

5–9. Answers will vary.

10. Answers will vary. The range is shown by the entire width of the box-and-whisker plot. It is the distance from the left whisker to the right whisker.
11. Answers will vary. The interquartile range is shown by the rectangle in the center of the box-and-whisker plot.
12. Answers will vary. Several combinations are possible. If both measures are equal, the data is evenly spread over the range. If the interquartile range is smaller than the range, the data has a small spread with outliers; streamflow was most stable over the decade with a few years of variant data. If the interquartile range is larger than the range, the data has a large spread; streamflow varied greatly over the decade.
13. Answers will vary. See possible answers to problem 12.

Differentiation

- Depending on available resources, students may create displays using paper, overhead materials, or computer software.
- Some students may benefit from the use of calculators during this task.
- Reduce the task load for some students by supplying a smaller set of numbers, or provide a set of numbers that are easier to place into a box-and-whisker plot. A data set with a total value divisible by 4 eases the calculation of the median and quartiles.
- Have students use technology to aid in the development of box-and-whisker plots.
- Allow students to verbally explain each of the components of a box-and-whisker plot.
- It may be helpful to provide a standard form for the creation of the box-and-whisker plot. This form could include: space for/instructions on the necessary calculations; space for the creation of the box-and-whisker plot; standardized units and/or placement of title and citation.

- Groups that finish early in the process should be encouraged to compare their box-and-whisker plot with another group. Use questions from the debriefing to encourage their between-group discussion. Have students explore how each individual decade compares to the larger time scale.
- Students or groups could be assigned the production of a box-and-whisker plot to summarize the entire data set. This provides another way to compare the data during the debriefing period.
- Have students predict what effect additional data could have when added to the set of data already represented by a box-and-whisker plot. Then create a new box-and-whisker plot that includes the new data and verify the prediction based on the new graphical representation.
- Advanced students can be encouraged to research and report on precipitation amounts over the decades studied. Students should investigate if precipitation data matched the conjectures developed in class.

Technology Connection

Have students use a graphing calculator or computer program to create box-and-whisker plots. Calculations necessary for the production of a box-and-whisker plot can be accomplished using the commands found in most spreadsheet programs. Using a spreadsheet program to calculate quartiles may lead to a discussion regarding the multiple methods used to find quartiles. You may wish to review the information about quartiles at the following link, for your own information before discussion.

www.walch.com/CCTTG6QuartilesInformation

Computer applets that are able to make box-and-whisker plots are available on many Web sites.

Choices for Students

Expansion of this task can be accomplished by incorporating other numeric data sets to complete and compare box-and-whisker plots. Additional data is available through USGS and the Chesapeake Bay Program (see Recommended Resources).

Meaningful Context

This task makes use of existing data from the Chesapeake Bay, a region of vital importance to Maryland. Additional data sets for this region are available online at the following site:

www.walch.com/rr/CCTTG6ChesapeakeBay

Recommended Resources

- Chesapeake Bay Program
  www.walch.com/rr/CCTTG6ChesapeakeBay
  This Web site provides historical information and data about the U.S.'s largest estuary, the Chesapeake Bay.
- Quartiles and Box-and-Whisker Plots
  www.walch.com/rr/CCTTG6Quartiles
  This site provides a description of what quartiles are and how to calculate them given a data set. It also includes directions for constructing box-and-whisker plots.
- USGS—U.S. Geological Survey
  www.walch.com/rr/CCTTG6USGeologicalSurvey
  This is the official site of the U.S. Geological Survey, an organization that conducts research on the environment, including climate change and ecosystem health.

NAME:

6.SP.4 Task • Statistics and Probability
Water Flowing into the Chesapeake Bay

Part 1

Use the "Estimated Annual-Mean Streamflow Entering Chesapeake Bay, By Water Year" data, found at www.walch.com/CCTTG6WaterData, to answer the following questions.

1. What information is being given in the data table?

2. Was this information collected or calculated?

3. What area does this data cover?

4. What time period does this data cover?

continued

6.SP.4 Task • Statistics and Probability
Water Flowing into the Chesapeake Bay

Part 2

Your group will work together to summarize 10 years of streamflow data. You will then present your data summary to the class. The data summary you create will be used to compare streamflow during each time period. Fill in the following information before you begin:

We are responsible for the data from year _______ to year _______

The title of our box and whisker plot: ____

Find the following information.

5. List the data values below.

6. Find the median of the data.

7. List the maximum and minimum values.

8. Find the upper and lower quartiles.

9. Use the materials given to create a box-and-whisker plot to summarize the data.

continued

6.SP.4 Task • Statistics and Probability
Water Flowing into the Chesapeake Bay

Part 3

Use the box-and-whisker plot created by your group to answer the following questions.

10. Find the range of your data. Describe how this value is shown on the box-and-whisker plot.

11. Find the interquartile range of your data. Describe how this value is shown on the box-and-whisker plot.

12. Half of your data can be found in the two whiskers of your box-and-whisker plot. Half of your data can be found in the interquartile range. Compare the length of the whiskers to the length of the interquartile range.

13. How does the average amount of water flowing into the Chesapeake Bay vary from year to year during the decade you were assigned?

How Tall Is a Sixth Grader?

Common Core State Standard

Summarize and describe distributions.

6.SP.5c. Summarize numerical data sets in relation to their context, such as by giving quantitative measure of center (median and/or mean) and variability (interquartile range and/or mean absolute deviation), as well as describing any overall pattern and any striking deviations from the overall pattern with reference to the context in which the data were gathered.

Task Overview

Background

The real world is full of applications in which one needs to look at data, summarize data by evaluating measures of center and variability, and recognize patterns or exceptions. Students today are bombarded with data and need to be skilled at being able to critically analyze it.

Students will collect data by measuring one another's heights, recording the measures, and then analyzing the data from the entire class by determining mean and median and creating a histogram. They will then look at what happens to the mean and median if some of the data are removed.

The task also provides practice with:

- measuring
- identifying outliers
- recognizing how adding or deleting data points affects measures of center

Implementation Suggestions

Students should be divided into pairs or groups of three. Students will need measuring tapes or yardsticks; calculators aren't absolutely necessary but will help save class time. It might be helpful to have a data sheet prepared in advance for students to use to record the heights (see sample below).

Student name	Height in inches	Student name	Height in inches

Introduction

The task should be introduced following instruction of quantitative measures of center and variability. Students should be able to calculate mean and median of a data set and should understand their role in describing and analyzing a data set. Students should be able to plot data in a histogram or bar graph.

Begin the task by having students recall how to find the median and mean of a data set. As a class, review the process of creating a histogram or bar graph. Discuss how to label the axes and determine a scale. Be prepared to demonstrate the process for calculating quartiles and interquartile range.

Make sure students know how to use the measuring tape or yardstick to accurately measure one another's heights in inches.

Discuss why it is important to measure the heights in inches instead of feet and inches (so the data can be worked with more easily).

Tell students they will be working in small groups. Explain that each student will measure another and each group will record their own measures. Have a place where students can record their data so the entire class can see—either on the board or on an overhead projector. Each student should record the data for the entire class on their own sheet. Each student should calculate the median and mean of the entire class data.

Group students in pairs or groups of three. Provide each group with the materials needed to make the measurements, record the data, and create their histograms based on the data for the entire class.

Monitoring/Facilitating the Task

Ask questions and prompt student thinking so that they:

- Accurately measure one another's height.
- Work so that each student has a chance to perform a measurement and record data.
- Correctly record the data for the entire class. Within each small group, each student should calculate the mean and median and then compare individual results with one another.
- Create a histogram based on the data for the entire class, not just for their group. Explain that a histogram needs several data points in order to be valid.
- Understand how to correctly calculate quartiles and interquartile range.
- Look for any data points that seem to be different from the rest and identify these data points mathematically as outliers.
- Think about what happens to the median and mean of a data set if some of the data points are removed. Consider especially what happens if any outliers are removed.

Debriefing the Task

- Upon completion of the task, students should share their results. Make sure all of the students have copied the measurements correctly. The histograms and means and medians for the entire class should reflect the same information but may be set up differently, or calculated/rounded slightly differently.
- Use this opportunity to look at variation and point out that even without looking exactly the same, the histograms and calculations should "paint the same picture" of the data.
- Ask each group to describe which data points they identified as outliers. Ask students to describe how they determined this. (Answers may vary depending on the original data set.)
- Ask each group to describe what happened when they removed some of the data points. Encourage students to articulate their answers using proper mathematical terms such as, "the mean or median increased or decreased," or "the revised mean or median is less than or greater than the original mean or median."

Answer Key

1. Answers will vary depending on the data collected; check for completion.
2. Answers will vary; check that students accurately recorded class data.
3. Answers will vary depending on the class data, but each group should arrive at the same answers for mean and median.
4. Histograms will depend on the class data, but should be identical or nearly identical across the groups, as they're all using the same data.

Answers for 5–11 will vary depending on the data collected. A sample data set and answers derived from it are provided below to use as a guide.

Sample data set: 50, 56, 58, 58, 59, 60, 60, 60, 61, 61, 62, 64

5. Students should identify any data points that appear to be significantly higher or lower than the rest of the data. For the sample data set, 50 and 64 seem to be different from the rest of the data.
6. Sample answer: $Q_1 = 58$ and $Q_3 = 61$; IQR = 61 – 58 = 3
7. Sample answer: 1.5 • 3 = 4.5; outliers would be less than 58 – 4.5 = 53.5 or greater than 61 + 4.5 = 65.5. Therefore, 50 is an outlier but 64 is not.
8. Sample answer: When the 50 is removed from the data, the median remains at 60 and the mean changes to 60, which is higher than the original mean of 59.
9. Sample answer: The mean of the original data set is 59. There is one data point equal to 59. When that data point is removed, the revised mean is still 59.

10. Sample answer: The median of the original data set is 60. There are three data points equal to 60. When one of the data points for 60 is removed, the revised median is still 60.
11. Sample answer: If there are outliers in a data set, removing them appears to change the mean of the data set. The median does not always appear to be affected. However, when a data point near the center of the data is removed, the median does not appear to change. It seems that removing outliers from data has more of an impact on measures of center than removing data points that are equal to or nearly equal to the median.

Differentiation

- Some students may benefit through the use of calculators during this task.
- It may be helpful to group students in such a way that students who may struggle with this task are paired with students who will likely not struggle.
- Allow students to use the "Plop It!" interactive Web tool listed in the Recommended Resources section to create the histograms and to calculate the mean and median.
- If possible, have someone fill in the class data for students for whom recording the data would prove difficult.
- Allow students to use computer software (Excel, for example) to record the data, calculate the mean and median, and create a histogram.
- Students who complete the task early could calculate the mean absolute deviation of the data and think about how that measurement helps describe the data.

Technology Connection

Students could use a graphing calculator or a spreadsheet program to compile the data, calculate the mean and median, and create a histogram. Students can also use the "Plop It!" tool to calculate the mean and median and create the histogram.

Choices for Students

Following the introduction, offer students the opportunity to calculate means and medians and create histograms using additional data. This data could include grades, birth dates, number of siblings, etc.

Meaningful Context

This task makes use of actual data that the students collect themselves. This task allows students to compile data, determine measures of center, and make conclusions about the data. The discussion of the data helps students to see how accurate or inaccurate statements about "the typical sixth grader's height" or "this class's average height" are.

Recommended Resources

- Illuminations—Histogram Tool
  www.walch.com/rr/CCTTG6HistogramTool
  This online tool enables students to create a histogram using data that they enter or by using pre-loaded data. Requires Java to run.
- Seeing Math—Plop It!
  www.walch.com/rr/CCTTG6PlopIt
  Students can use this interactive software to highlight how changing a data set affects the mean, median, and mode. Requires Java to run.

6.SP.5(c) Task • Statistics and Probability
How Tall Is a Sixth Grader?

Part 1

How tall is a "typical" sixth grader? How tall are the students in your math class? In this task, you are going to use actual height measurements of the students in your class to collect and analyze data. First, you will measure one another, collect all of the data for the class, and create a histogram of the data. You will work in pairs or groups of three, but each of you should complete a data sheet and histogram.

1. In your group, measure one another in inches and record the data. Round to the nearest inch and use the data sheet provided by your teacher.

2. Write the data on the class record and copy all of the other groups' data on your own data sheet. Make sure the data has been rounded to the nearest inch.

3. Calculate the mean and median for the data set. Round to the nearest inch. Describe how the mean and median of the data set relate to each other.

4. Create a histogram of the data.

continued

6.SP.5(c) Task • Statistics and Probability
How Tall Is a Sixth Grader?

Part 2

Now you will analyze the data from Part 1 by determining measures of center (mean and median) and the measure of variability (interquartile range, or IQR). You will also look at what happens to the measures of center if the data set is changed. This will help you to make statements about how tall a "typical" sixth grader is, or how tall students in your math class are.

5. Look at the data. Are there any data points that seem to be different from the others? How are they different?

6. Find the quartiles and the interquartile range (IQR). The first quartile (Q_1) is the median of the lower half of the data. The third quartile (Q_3) is the median of the upper half of the data. Find the IQR by subtracting Q_1 from Q_3 ($Q_3 - Q_1$).

7. Determine if there are any outliers. Outliers are data points that are more than Q_3 plus 1.5 times the IQR or less than Q_1 minus 1.5 times the IQR.

   List any outliers you found: ____________________

continued

8. What happens to the mean and median if you remove any outliers from your calculations? Describe this fully.

9. Looking at the original data set, find one piece of data that is equal to (or nearly equal to) the mean. How does the mean change if you remove this piece of data from your calculations?

10. Looking at the original data set, find one piece of data that is equal to (or nearly equal to) the median. How does the median change if you remove this piece of data from your calculations?

11. Think about what happened to the mean and median when you removed an outlier from the calculations. What happened to the measures of center when you removed a piece of data that was equal to (or nearly equal to) the mean or median? What do you think happens to the measures of center when you modify your data set? Explain below.